LINEAR INTEGRAL EQUATIONS
Theory and Technique

Linear
Integral Equations
Theory and Technique

RAM P. KANWAL

Pennsylvania State University

University Park, Pennsylvania

ACADEMIC PRESS   **1971**   New York and London

ACADEMIC PRESS, INC.
111 Fifth Avenue, New York, New York 10003

United Kingdom Edition published by
ACADEMIC PRESS, INC. (LONDON) LTD.
Berkeley Square House, London W1X 6BA

LIBRARY OF CONGRESS CATALOG CARD NUMBER: 77-156268

AMS (MOS) 1970 Subject Classification: 45A05, 45B05, 45C05,
45D05, 45E05, 45E10, 45E99, 45F05, 35C15, 44A25, 44A35, 47G05

PRINTED IN THE UNITED STATES OF AMERICA

IN LOVING MEMORY OF
MY GRANDMOTHER

CONTENTS

PREFACE

Many physical problems which are usually solved by differential equation methods can be solved more effectively by integral equation methods. Indeed, the latter have been appearing in current literature with increasing frequency and have provided solutions to problems heretofore not solvable by standard methods of differential equations. Such problems abound in many applied fields, and the types of solutions explored here will be useful particularly in applied mathematics, theoretical mechanics, and mathematical physics.

Each section of the book contains a selection of examples based on the technique of that section, thus making it possible to use the book as the text for a beginning graduate course. The latter part of the book will be equally useful to research workers who encounter boundary value problems. The level of mathematical knowledge required of the reader is no more than that taught in undergraduate applied mathematics courses. Although no attempt has been made to attain a high level of mathematical rigor, the regularity conditions for every theorem have been stated precisely. To keep the book to a manageable size, a few long proofs have been omitted. They are mostly those proofs which do not appear to be essential in the study of the subject for purposes of applications.

We have omitted topics such as Wiener–Hopf technique and dual integral equations because most of the problems which can be solved by these methods can be solved more easily by the integral equation methods presented in Chapters 10 and 11. Furthermore, we have concentrated mainly on three-dimensional problems. Once the methods have been grasped, the student can readily solve the corresponding plane problems. Besides, the powerful tools of complex variables are available for the latter problems.

The book has developed from courses of lectures on the subject given by the author over a period of years at Pennsylvania State University.

Since it is intended mainly as a textbook we have omitted references to the current research literature. The bibliography contains references to books which the author has consulted and to which the reader is referred occasionally in the text. Chapter 10 is based on author's joint article with Dr. D. L. Jain (*SIAM Journal of Applied Mathematics*, **20**, 1971) while Chapter 11 is based on an article by the author (*Journal of Mathematics and Mechanics*, **19**, 1970, 625–656). The permission of the publishers of these two journals is thankfully acknowledged. These articles contain references to the current research literature pertaining to these chapters. The most important of these references are the research works of Professor W. E. Williams and Professor B. Noble.

Finally, I would like to express my gratitude to my students Mr. P. Gressis, Mr. A. S. Ibrahim, Dr. D. L. Jain, and Mr. B. K. Sachdeva for their assistance in the preparation of the textual material, to Mrs. Suzie Mostoller for her pertinent typing of the manuscript, and to the staff of Academic Press for their helpful cooperation.

LINEAR INTEGRAL EQUATIONS
Theory and Technique

INTRODUCTION

1.1. DEFINITION

An integral equation is an equation in which an unknown function appears under one or more integral signs. Naturally, in such an equation there can occur other terms as well. For example, for $a \leqslant s \leqslant b$, $a \leqslant t \leqslant b$, the equations

$$f(s) = \int_a^b K(s,t)\,g(t)\,dt \, , \tag{1}$$

$$g(s) = f(s) + \int_a^b K(s,t)\,g(t)\,dt \, , \tag{2}$$

$$g(s) = \int_a^b K(s,t)\,[g(t)]^2\,dt \, , \tag{3}$$

where the function $g(s)$ is the unknown function while all the other functions are known, are integral equations. These functions may be complex-valued functions of the real variables s and t.

Integral equations occur naturally in many fields of mechanics and mathematical physics. They also arise as representation formulas for the solutions of differential equations. Indeed, a differential equation can be

1

replaced by an integral equation which incorporates its boundary conditions. As such, each solution of the integral equation automatically satisfies these boundary conditions. Integral equations also form one of the most useful tools in many branches of pure analysis, such as the theories of functional analysis and stochastic processes.

One can also consider integral equations in which the unknown function is dependent not only on one variable but on several variables. Such, for example, is the equation

$$g(s) = f(s) + \int_\Omega K(s, t) g(t) \, dt \,, \tag{4}$$

where s and t are n-dimensional vectors and Ω is a region of an n-dimensional space. Similarly, one can also consider systems of integral equations with several unknown functions.

An integral equation is called linear if only linear operations are performed in it upon the unknown function. The equations (1) and (2) are linear, while (3) is nonlinear. In fact, the equations (1) and (2) can be written as

$$L[g(s)] = f(s) \,, \tag{5}$$

where L is the appropriate integral operator. Then, for any constants c_1 and c_2, we have

$$L[c_1 g_1(s) + c_2 g_2(s)] = c_1 L[g_1(s)] + c_2 L[g_2(s)] \,. \tag{6}$$

This is the general criterion for a linear operator. In this book, we shall deal only with linear integral equations.

The most general type of linear integral equation is of the form

$$h(s) g(s) = f(s) + \lambda \int_a K(s, t) g(t) \, dt \,, \tag{7}$$

where the upper limit may be either variable or fixed. The functions f, h, and K are known functions, while g is to be determined; λ is a nonzero, real or complex, parameter. The function $K(s, t)$ is called the kernel. The following special cases of equation (7) are of main interest.

(i)   FREDHOLM INTEGRAL EQUATIONS.   In all Fredholm integral equations, the upper limit of integration b, say, is fixed.

   (i)   In the Fredholm integral equation of the first kind, $h(s) = 0$.

Thus,

$$f(s) + \lambda \int_a^b K(s,t) g(t) \, dt = 0 \, . \tag{8}$$

(ii)   In the Fredholm integral equation of the second kind, $h(s) = 1$;

$$g(s) = f(s) + \lambda \int_a^b K(s,t) g(t) \, dt \, . \tag{9}$$

(iii)   The homogeneous Fredholm integral equation of the second kind is a special case of (ii) above. In this case, $f(s) = 0$;

$$g(s) = \lambda \int_a^b K(s,t) g(t) \, dt \, . \tag{10}$$

(ii)   VOLTERRA EQUATIONS.   Volterra equations of the first, homogeneous, and second kinds are defined precisely as above except that $b = s$ is the variable upper limit of integration.

Equation (7) itself is called an integral equation of the third kind.

(iii)   SINGULAR INTEGRAL EQUATIONS. When one or both limits of integration become infinite or when the kernel becomes infinite at one or more points within the range of integration, the integral equation is called singular. For example, the integral equations

$$g(s) = f(s) + \lambda \int_{-\infty}^{\infty} (\exp - |s-t|) g(t) \, dt \tag{11}$$

and

$$f(s) = \int_0^s [1/(s-t)^\alpha] g(t) \, dt \, , \qquad 0 < \alpha < 1 \tag{12}$$

are singular integral equations.

1.2. REGULARITY CONDITIONS

We shall be mainly concerned with functions which are either continuous, or integrable or square-integrable. In the subsequent analysis, it will be pointed out what regularity conditions are expected of the functions involved. The notion of Lebesgue-integration is essential to modern mathematical analysis. When an integral sign is used, the Lebesgue integral is understood. Fortunately, if a function is Riemann-integrable, it is also Lebesgue-integrable. There are functions that are Lebesgue-integrable but not Riemann-integrable, but we shall not encounter them in this book. Incidentally, by a square-integrable function $g(t)$, we mean that

$$\int_a^b |g(t)|^2 \, dt < \infty .\tag{1}$$

This is called an $\mathscr{L}_2$-function. The regularity conditions on the kernel $K(s, t)$ as a function of two variables are similar. It is an $\mathscr{L}_2$-function if:

(a) for each set of values of s, t in the square $a \leqslant s \leqslant b \; a \leqslant t \leqslant b$,

$$\int_a^b\int_a^b |K(s, t)|^2 \, ds \, dt < \infty ,\tag{2}$$

(b) for each value of s in $a \leqslant s \leqslant b$,

$$\int_a^b |K(s, t)|^2 \, dt < \infty ,\tag{3}$$

(c) for each value of t in $a \leqslant t \leqslant b$,

$$\int_a^b |K(s, t)|^2 \, ds < \infty .\tag{4}$$

1.3. SPECIAL KINDS OF KERNELS

(i) SEPARABLE OR DEGENERATE KERNEL. A kernel $K(s, t)$ is called separable or degenerate if it can be expressed as the sum of a finite number of terms, each of which is the product of a function of s only and a function of t only, i.e.,

$$K(s, t) = \sum_{i=1}^n a_i(s) b_i(t) .\tag{1}$$

The functions $a_i(s)$ can be assumed to be linearly independent, otherwise the number of terms in relation (1) can be reduced (by linear independence it is meant that, if $c_1 a_1 + c_2 a_2 + \cdots + c_n a_n = 0$, where c_i are arbitrary constants, then $c_1 = c_2 = \cdots = c_n = 0$).

(ii) SYMMETRIC KERNEL. A complex-valued function $K(s,t)$ is called symmetric (or Hermitian) if $K(s,t) = K^*(t,s)$, where the asterisk denotes the complex conjugate. For a real kernel, this coincides with definition $K(s,t) = K(t,s)$.

1.4. EIGENVALUES AND EIGENFUNCTIONS

If we write the homogeneous Fredholm equation as

$$\int_a^b K(s,t) g(t)\, dt = \mu g(s), \qquad \mu = 1/\lambda,$$

we have the classical eigenvalue or characteristic value problem; μ is the eigenvalue and $g(t)$ is the corresponding eigenfunction or characteristic function. Since the linear integral equations are studied in the form (1.1.10), it is λ and not $1/\lambda$ which is called the eigenvalue.

1.5. CONVOLUTION INTEGRAL

Many interesting problems of mechanics and physics lead to an integral equation in which the kernel $K(s,t)$ is a function of the difference $(s-t)$ only:

$$K(s,t) = k(s-t), \qquad (1)$$

where k is a certain function of one variable. The integral equation

$$g(s) = f(s) + \lambda \int_a^s k(s-t) g(t)\, dt, \qquad (2)$$

and the corresponding Fredholm equation are called integral equations of the convolution type.

The function defined by the integral

$$\int_0^s k(s-t)\,g(t)\,dt = \int_0^s k(t)\,g(s-t)\,dt \tag{3}$$

is called the convolution or the Faltung of the two functions k and g. The integrals occurring in (3) are called the convolution integrals.

The convolution defined by relation (3) is a special case of the standard convolution

$$\int_{-\infty}^{\infty} k(s-t)\,g(t)\,dt = \int_{-\infty}^{\infty} k(t)\,g(s-t)\,dt\,. \tag{4}$$

The integrals in (3) are obtained from those in (4) by taking $k(t) = g(t) = 0$, for $t < 0$ and $t > s$.

1.6. THE INNER OR SCALAR PRODUCT OF TWO FUNCTIONS

The inner or scalar product (ϕ, ψ) of two complex $\mathcal{L}_2$-functions ϕ and ψ of a real variable s, $a \leqslant s \leqslant b$, is defined as

$$(\phi, \psi) = \int_a^b \phi(t)\,\psi^*(t)\,dt\,. \tag{1}$$

Two functions are called orthogonal if their inner product is zero, that is, ϕ and ψ are orthogonal if $(\phi, \psi) = 0$. The norm of a function $\phi(t)$ is given by the relation

$$\|\phi\| = \left[\int_a^b \phi(t)\,\phi^*(t)\,dt\right]^{\frac{1}{2}} = \left[\int_a^b |\phi(t)|^2\,dt\right]^{\frac{1}{2}}\,. \tag{2}$$

A function ϕ is called normalized if $\|\phi\| = 1$. It is obvious that a nonnull function (whose norm is not zero) can always be normalized by dividing it by its norm.

We shall have a great deal more to say about these ideas in Chapter 7. For the time being, we shall content ourselves with mentioning the Schwarz and Minkowski inequalities,

$$|(\phi, \psi)| \leqslant \|\phi\| \|\psi\| \tag{3}$$

and

$$\|\phi + \psi\| \leqslant \|\phi\| + \|\psi\|, \tag{4}$$

respectively.

1.7. NOTATION

For Fredholm integral equations, it will be assumed that the range of integrations is a to b, unless the contrary is explicitly stated. The quantities a and b will be omitted from the integral sign in the sequel.

INTEGRAL EQUATIONS WITH SEPARABLE KERNELS

2.1. REDUCTION TO A SYSTEM OF ALGEBRAIC EQUATIONS

In Chapter 1, we have defined a degenerate or a separable kernel $K(s, t)$ as

$$K(s, t) = \sum_{i=1}^{n} a_i(s) b_i(t) , \tag{1}$$

where the functions $a_1(s), ..., a_n(s)$ and the functions $b_1(t), ..., b_n(t)$ are linearly independent. With such a kernel, the Fredholm integral equation of the second kind,

$$g(s) = f(s) + \lambda \int K(s, t) g(t) dt \tag{2}$$

becomes

$$g(s) = f(s) + \lambda \sum_{i=1}^{n} a_i(s) \int b_i(t) g(t) dt . \tag{3}$$

It emerges that the technique of solving this equation is essentially dependent on the choice of the complex parameter λ and on the definition of

$$c_i = \int b_i(t) g(t) dt . \tag{4}$$

The quantities c_i are constants, although hitherto unknown.

Substituting (4) in (3) gives

$$g(s) = f(s) + \lambda \sum_{i=1}^{n} c_i a_i(s) , \tag{5}$$

and the problem reduces to finding the quantities c_i. To this end, we put the value of $g(s)$ as given by (5) in (3) and get

$$\sum_{i=1}^{n} a_i(s) \{c_i - \int b_i(t) [f(t) + \lambda \sum_{k=1}^{n} c_k a_k(t)] \, dt\} = 0 . \tag{6}$$

But the functions $a_i(s)$ are linearly independent; therefore,

$$c_i - \int b_i(t) [f(t) + \lambda \sum_{k=1}^{n} c_k a_k(t)] \, dt = 0 , \qquad i = 1,...,n . \tag{7}$$

Using the simplified notation

$$\int b_i(t) f(t) = f_i , \qquad \int b_i(t) a_k(t) \, dt = a_{ik} , \tag{8}$$

where f_i and a_{ik} are known constants, equation (7) becomes

$$c_i - \lambda \sum_{k=1}^{n} a_{ik} c_k = f_i , \qquad i = 1,...,n ; \tag{9}$$

that is, a system of n algebraic equations for the unknowns c_i. The determinant $D(\lambda)$ of this system is

$$D(\lambda) = \begin{vmatrix} 1 - \lambda a_{11} & -\lambda a_{12} & \cdots & -\lambda a_{1n} \\ -\lambda a_{21} & 1 - \lambda a_{22} & \cdots & -\lambda a_{2n} \\ \vdots & & & \\ -\lambda a_{n1} & -\lambda a_{n2} & \cdots & 1 - \lambda a_{nn} \end{vmatrix} , \tag{10}$$

which is a polynomial in λ of degree at most n. Moreover, it is not identically zero, since, when $\lambda = 0$, it reduces to unity.

For all values of λ for which $D(\lambda) \neq 0$, the algebraic system (9), and thereby the integral equation (2), has a unique solution. On the other hand, for all values of λ for which $D(\lambda)$ becomes equal to zero, the algerbaic system (9), and with it the integral equation (2), either is insoluble or has an infinite number of solutions. Setting $\lambda = 1/\mu$ in equation (9), we have the eigenvalue problem of matrix theory. The eigenvalues are given by the polynomial $D(\lambda) = 0$. They are also the eigenvalues of our integral equation.

Note that we have considered only the integral equation of the second kind, where alone this method is applicable.

This method is illustrated with the following examples.

2.2. EXAMPLES

Example 1.   Solve the Fredholm integral equation of the second kind

$$g(s) = s + \lambda \int_0^1 (st^2 + s^2 t)g(t)\, dt . \tag{1}$$

The kernel $K(s, t) = st^2 + s^2 t$ is separable and we can set

$$c_1 = \int_0^1 t^2 g(t)\, dt , \qquad c_2 = \int_0^1 tg(t)\, dt .$$

Equation (1) becomes

$$g(s) = s + \lambda c_1 s + \lambda c_2 s^2 , \tag{2}$$

which we substitute in (1) to obtain the algebraic equations

$$\begin{aligned}
c_1 &= \tfrac{1}{4} + \tfrac{1}{4}\lambda c_1 + \tfrac{1}{5}\lambda c_2 , \\
c_2 &= \tfrac{1}{3} + \tfrac{1}{3}\lambda c_1 + \tfrac{1}{4}\lambda c_2 .
\end{aligned} \tag{3}$$

The solution of these equations is readily obtained as

$$c_1 = (60 + \lambda)/(240 - 120\lambda - \lambda^2) , \qquad c_2 = 80/(240 - 120\lambda - \lambda^2) . \tag{4}$$

From (2) and (4), we have the solution

$$g(s) = [(240 - 60\lambda)s + 80\lambda s^2]/(240 - 120\lambda - \lambda^2) . \tag{5}$$

Example 2.   Solve the integral equation

$$g(s) = f(s) + \lambda \int_0^1 (s + t)g(t)\, dt \tag{6}$$

and find the eigenvalues.

Here, $a_1(s) = s$, $a_2(s) = 1$, $b_1(t) = 1$, $b_2(t) = t$,

$$a_{11} = \int_0^1 t\, dt = \tfrac{1}{2}, \qquad a_{12} = \int_0^1 dt = 1,$$

$$a_{21} = \int_0^1 t^2\, dt = \tfrac{1}{3}, \qquad a_{22} = \int_0^1 t\, dt = \tfrac{1}{2},$$

$$f_1 = \int_0^1 f(t)\, dt, \qquad f_2 = \int_0^1 tf(t)\, dt.$$

Substituting these values in (2.1.9), we have the algebraic system

$$(1 - \tfrac{1}{2}\lambda)c_1 - \lambda c_2 = f_1, \qquad -\tfrac{1}{3}\lambda c_1 + (1 - \tfrac{1}{2}\lambda)c_2 = f_2.$$

The determinant $D(\lambda) = 0$ gives $\lambda^2 + 12\lambda - 12 = 0$. Thus, the eigenvalues are

$$\lambda_1 = (-6 + 4\sqrt{3}), \qquad \lambda_2 = (-6 - 4\sqrt{3}).$$

For these two values of λ, the homogeneous equation has a nontrivial solution, while the integral equation (6) is, in general, not soluble. When λ differs from these values, the solution of the above algebraic system is

$$c_1 = [-12f_1 + \lambda(6f_1 - 12f_2)]/(\lambda^2 + 12\lambda - 12),$$

$$c_2 = [-12f_2 - \lambda(4f_1 - 6f_2)]/(\lambda^2 + 12\lambda - 12).$$

Using the relation (2.1.5), there results the solution

$$g(s) = f(s) + \lambda \int_0^1 \frac{6(\lambda - 2)(s+t) - 12\lambda st - 4\lambda}{\lambda^2 + 12\lambda - 12} f(t)\, dt. \tag{7}$$

The function $\Gamma(s, t; \lambda)$,

$$\Gamma(s, t; \lambda) = [6(\lambda - 2)(s+t) - 12\lambda st - 4\lambda]/(\lambda^2 + 12\lambda - 12), \tag{8}$$

is called the resolvent kernel. We have therefore succeeded in inverting the integral equation because the right-hand side of the above formula is a known quantity.

Example 3. Invert the integral equation

$$g(s) = f(s) + \lambda \int_0^{2\pi} (\sin s \cos t) g(t) \, dt \, . \tag{9}$$

As in the previous examples, we set

$$c = \int_0^{2\pi} (\cos t) g(t) \, dt$$

to obtain

$$g(s) = f(s) + \lambda c \sin s \, . \tag{10}$$

Multiply both sides of this equation by $\cos s$ and integrate from 0 to 2π. This gives

$$c = \int_0^{2\pi} (\cos t) f(t) \, dt \, . \tag{11}$$

From (10) and (11), we have the required formula:

$$g(s) = f(s) + \lambda \int_0^{2\pi} (\sin s \cos t) f(t) \, dt \, . \tag{12}$$

Example 4. Find the resolvent kernel for the integral equation

$$g(s) = f(s) + \lambda \int_{-1}^{1} (st + s^2 t^2) g(t) \, dt \, . \tag{13}$$

For this equation,

$$a_1(s) = s \, , \qquad a_2(s) = s^2 \, , \qquad b_1(t) = t \, , \qquad b_2(t) = t^2 \, ,$$

$$a_{11} = \tfrac{2}{3} \, , \qquad a_{12} = a_{21} = 0 \, , \qquad a_{22} = \tfrac{2}{5} \, ,$$

$$f_1 = \int_{-1}^{1} t f(t) \, dt \, , \qquad f_2 = \int_{-1}^{1} t^2 f(t) \, dt \, .$$

Therefore, the corresponding algebraic system is

$$c_1(1 - \tfrac{2}{3}\lambda) = f_1 \, , \qquad c_2(1 - \tfrac{2}{5}\lambda) = f_2 \, . \tag{14}$$

Substituting the values of c_1 and c_2 as obtained from (14) in (2.1.9) yields the solution

$$g(s) = f(s) + \lambda \int_{-1}^{1} \left[\frac{st}{(1-\frac{2}{3}\lambda)} + \frac{s^2 t^2}{(1-\frac{2}{5}\lambda)} \right] f(t) \, dt \ . \tag{15}$$

Thus, the resolvent kernel is

$$\Gamma(s, t; \lambda) = \frac{st}{1-(2\lambda/3)} + \frac{s^2 t^2}{1-(2\lambda/5)} \ . \tag{16}$$

We shall now give examples of homogeneous integral equations.

Example 5. Solve the homogeneous Fredholm integral equation

$$g(s) = \lambda \int_{0}^{1} e^s e^t g(t) \, dt \ . \tag{17}$$

Define

$$c = \int_{0}^{1} e^t g(t) \, dt \ ,$$

so that (17) becomes

$$g(s) = \lambda c e^s \ . \tag{18}$$

Put this value of $g(s)$ in (17) and get

$$\lambda c e^s = \lambda e^s \int_{0}^{1} e^t [\lambda c e^t] \, dt = \tfrac{1}{2}\lambda^2 e^s c(e^2 - 1)$$

or

$$\lambda c \{2 - \lambda(e^2 - 1)\} = 0 \ .$$

If $c = 0$ or $\lambda = 0$, then we find that $g \equiv 0$. Assume that neither $c = 0$ nor $\lambda = 0$; then, we have the eigenvalue

$$\lambda = 2/(e^2 - 1) \ .$$

Only for this value of λ does there exist a nontrivial solution of the integral equation (17). This solution is found from (18) to be

$$g(s) = [2c/(e^2 - 1)] e^s \ . \tag{19}$$

Thus, to the eigenvalue $2/(e^2 - 1)$ there corresponds the eigenfunction e^s.

Example 6. Find the eigenvalues and eigenfunctions of the homogeneous integral equation

$$g(s) = \lambda \int_1^2 [st + (1/st)] \, g(t) \, dt \; . \tag{20}$$

Comparing this with (2.1.3), we have

$$a_1(s) = s \, , \qquad a_2(s) = 1/s \, , \qquad b_1(t) = t \, , \qquad b_2(t) = 1/t \; .$$

$$a_{11} = \tfrac{7}{3} \, , \qquad a_{12} = a_{21} = 1 \, , \qquad a_{22} = \tfrac{1}{2} \; .$$

The formula (2.1.9) then yields the following homogeneous equations:

$$(1 - \tfrac{7}{3}\lambda) c_1 - \lambda c_2 = 0 \, , \qquad -\lambda c_1 + (1 - \tfrac{1}{2}\lambda) c_2 = 0 \, , \tag{21}$$

which have a nontrivial solution only if the determinant

$$D(\lambda) = \begin{vmatrix} 1 - \tfrac{7}{3}\lambda & -\lambda \\ -\lambda & 1 - \lambda \end{vmatrix} = 1 - (17/6)\lambda + (1/6)\lambda^2 \, ,$$

vanishes. Therefore, the required eigenvalues are

$$\begin{aligned} \lambda_1 &= \tfrac{1}{2}(17 + \sqrt{265}) \simeq 16.6394 \; ; \\ \lambda_2 &= \tfrac{1}{2}(17 - \sqrt{265}) \simeq 0.3606 \; . \end{aligned} \tag{22}$$

The solution corresponding to λ_1 is $c_2 \simeq -2.2732 c_1$, while that corresponding to λ_2 is $c_2' \simeq 0.4399 c_1'$. The eigenfunctions of the given integral equation are found by substituting in (2.1.5):

$$\begin{aligned} g_1(s) &\simeq 16.639 c_1 \, [s - 2.2732(1/s)] \, , \\ g_2(s) &\simeq 0.3606 c_1' \, [s + 0.4399(1/s)] \, , \end{aligned} \tag{23}$$

where c_1 and c_1' are two undetermined constants.

2.3. FREDHOLM ALTERNATIVE

In the previous sections, we have found that, if the kernel is separable, the problem of solving an integral equation of the second kind reduces to that of solving an algebraic system of equations. Unfortunately, integral equations with degenerate kernels do not occur frequently in practice. But since they are easily treated and, furthermore, the results derived for such equations lead to a better understanding of integral equations of more general types, it is worthwhile to study them. Last, but not least, any reasonably well-behaved kernel can be written as an infinite series of degenerate kernels.

When an integral equation cannot be solved in closed form, then recourse has to be taken to approximate methods. But these approximate methods can be applied with confidence only if the existence of the solution is assured in advance. The Fredholm theorems explained in this chapter provide such an assurance. The basic theorems of the general theory of integral equations, which were first presented by Fredholm, correspond to the basic theorems of linear algebraic systems. Fredholm's classical theory shall be presented in Chapter 4 for general kernels. Here, we shall deal with degenerate kernels and borrow the results of linear algebra.

In Section 2.1, we have found that the solution of the present problem rests on the investigation of the determinant (2.1.10) of the coefficients of the algebraic system (2.1.9). If $D(\lambda) \neq 0$, then that system has only one solution, given by Cramer's rule

$$c_i = (D_{1i}f_1 + D_{2i}f_2 + \cdots + D_{ni}f_n)/D(\lambda), \qquad i = 1, 2, \cdots, n, \quad (1)$$

where D_{hi} denotes the cofactor of the (h, i)th element of the determinant (2.1.10). Consequently, the integral equation (2.1.2) has the unique solution (2.1.5), which, in view of (1), becomes

$$g(s) = f(s) + \lambda \sum_{i=1}^{n} \frac{D_{1i}f_1 + D_{2i}f_2 + \cdots + D_{ni}f_n}{D(\lambda)} a_i(s), \qquad (2)$$

while the corresponding homogeneous equation

$$g(s) = \lambda \int K(s, t) g(t) \, dt \qquad (3)$$

has only the trivial solution $g(s) = 0$.

Substituting for f_i from (2.1.8) in (2), we can write the solution $g(s)$ as

$$g(s) = f(s) + [\lambda/D(\lambda)]$$

$$\times \int \left\{ \sum_{i=1}^{n} [D_{1i}b_1(t) + D_{2i}b_2(t) + \cdots + D_{ni}b_n(t)] a_i(s) \right\} f(t) \, dt.$$

$$(4)$$

Now consider the determinant of $(n+1)$th order

$$D(s,t;\lambda) = - \begin{vmatrix} 0 & a_1(s) & a_2(s) & \cdots & a_n(s) \\ b_1(t) & 1-\lambda a_{11} & -\lambda a_{12} & \cdots & -\lambda a_{1n} \\ b_2(t) & -\lambda a_{21} & 1-\lambda a_{22} & \cdots & -\lambda a_{2n} \\ \vdots & & & & \\ b_n(t) & -\lambda a_{n1} & -\lambda a_{n2} & \cdots & 1-\lambda a_{nn} \end{vmatrix}. \quad (5)$$

By developing it by the elements of the first row and the corresponding minors by the elements of the first column, we find that the expression in the brackets in equation (4) is $D(s,t;\lambda)$. With the definition

$$\Gamma(s,t;\lambda) = D(s,t;\lambda)/D(\lambda), \quad (6)$$

equation (4) takes the simple form

$$g(s) = f(s) + \lambda \int \Gamma(s,t;\lambda)f(t)\, dt. \quad (7)$$

The function $\Gamma(s,t;\lambda)$ is the resolvent (or reciprocal) kernel we have already encountered in Examples 2 and 4 in the previous section. We shall see in Chapter 4 that the formula (6) has many important consequences. For the time being, we content ourselves with the observation that the only possible singular points of $\Gamma(s,t;\lambda)$ in the λ-plane are the roots of the equation $D(\lambda) = 0$, i.e., the eigenvalues of the kernel $K(s,t)$.

The above discussion leads to the following basic Fredholm theorem.

Fredholm Theorem. The inhomogeneous Fredholm integral equation (2.1.2) with a separable kernel has one and only one solution, given by formula (7). The resolvent kernel $\Gamma(s,t;\lambda)$ coincides with the quotient (6) of two polynomials.

If $D(\lambda) = 0$, then the inhomogeneous equation (2.1.2) has no solution in general, because an algebraic system with vanishing determinant can be solved only for some particular values of the quantities f_i. To discuss this case, we write the algebraic system (2.1.9) as

$$(\mathbf{I}-\lambda\mathbf{A})\mathbf{c} = \mathbf{f}, \quad (8)$$

where $\mathbf{I}$ is the unit (or identity) matrix of order n and $\mathbf{A}$ is the matrix (a_{ij}). Now, when $D(\lambda) = 0$, we observe that for each nontrivial solution of the homogeneous algebraic system

$$(\mathbf{I}-\lambda\mathbf{A})\mathbf{c} = 0 \quad (9)$$

there corresponds a nontrivial solution (an eigenfunction) of the homogeneous integral equation (3). Furthermore, if λ coincides with a certain eigenvalue λ_0 for which the determinant $D(\lambda_0) = |\mathbf{I} - \lambda_0\mathbf{A}|$ has the rank p, $1 \leqslant p \leqslant n$, then there are $r = n - p$ linearly independent solutions of the algebraic system (9); r is called the index of the eigenvalue λ_0. The same holds for the homogeneous integral equation (3). Let us denote these r linearly independent solutions as $g_{01}(s), g_{02}(s), \cdots, g_{0r}(s)$, and let us also assume that they have been normalized. Then, to each eigenvalue λ_0 of index $r = n - p$, there corresponds a solution $g_0(s)$ of the homogeneous integral equation (3) of the form

$$g_0(s) = \sum_{k=1}^{r} \alpha_k g_{0k}(s) ,$$

where α_k are arbitrary constants.

Let m be the multiplicity of the eigenvalue λ_0, i.e., $D(\lambda) = 0$ has m equal roots λ_0. Then, we infer from the theory of linear algebra that, by using the elementary transformations on the determinant $|\mathbf{I} - \lambda\mathbf{A}|$, we shall have at most $m + 1$ identical rows and this maximum is achieved only if $\mathbf{A}$ is symmetric. This means that the rank p of $D(\lambda_0)$ is greater than or equal to $n - m$. Thus,

$$r = n - p \leqslant n - (n - m) = m ,$$

and the equality holds only when $a_{ij} = a_{ji}$.

Thus we have proved the theorem of Fredholm that, if $\lambda = \lambda_0$ is a root of multiplicity $m \geqslant 1$ of the equation $D(\lambda) = 0$, then the homogeneous integral equation (3) has r linearly independent solutions; r is the index of the eigenvalue such that $1 \leqslant r \leqslant m$.

The numbers r and m are also called the geometric multiplicity and algebraic multiplicity of λ_0, respectively. From the above result, it follows that the algebraic multiplicity of an eigenvalue must be greater than or equal to its geometric multiplicity.

To study the case when the inhomogeneous Fredholm integral equation (2.1.2) has solutions even when $D(\lambda) = 0$, we need to define and study the transpose of the equation (2.1.2). The integral equation[1]

$$\psi(s) = f(s) + \lambda \int K(t, s)\psi(t)\,dt \qquad (10)$$

---

[1] We shall consider only real functions here even though the results are easily extended to complex functions. Outcome is the same in both the cases.

is called the transpose (or adjoint) of the equation (2.1.2). Observe that the relation between (2.1.2) and its transpose (10) is symmetric, since (2.1.2) is the transpose of (10).

If the separable kernel $K(s, t)$ has the expansion (2.1.1), then the kernel $K(t, s)$ of the transposed equation has the expansion

$$K(t, s) = \sum_{i=1}^{n} a_i(t) b_i(s) . \tag{11}$$

Proceeding as in Section 2.1, we end up with the algebraic system

$$(\mathbf{I} - \lambda \mathbf{A}^{\mathrm{T}}) \mathbf{c} = \mathbf{f} , \tag{12}$$

where $\mathbf{A}^{\mathrm{T}}$ stands for the transpose of $\mathbf{A}$ and where c_i and f_i are now defined by the relations

$$c_i = \int a_i(t) g(t) \, dt , \qquad f_i = \int a_i(t) f(t) \, dt . \tag{13}$$

The interesting feature of the system (12) is that the determinant $D(\lambda)$ is the same function as (2.1.10) except that there has been an interchange of rows and columns in view of the interchange in the functions a_i and b_i. Thus, the eigenvalues of the transposed integral equation are the same as those of the original equation. This means that *the transposed equation* (10) *also possesses a unique solution whenever* (2.1.2) *does*.

As regards the eigenfunctions of the homogeneous system

$$|\mathbf{I} - \lambda \mathbf{A}^{\mathrm{T}}| \mathbf{c} = 0 , \tag{14}$$

we know from linear algebra that these are different from the corresponding eigenfunctions of the system (9). The same applies to the eigenfunctions of the transposed integral equation. Since the index r of λ_0 is the same in both these systems, the number of linearly independent eigenfunctions is also r for the transposed system. Let us denote them by $\psi_{01}, \psi_{02}, \cdots, \psi_{0r}$ and let us assume that they have been normalized. Then, any solution $\psi_0(s)$ of the transposed homogeneous integral equation

$$\psi(s) = \lambda \int K(t, s) \psi(t) \, dt \tag{15}$$

corresponding to the eigenvalue λ_0 is of the form

$$\psi_0(s) = \sum \beta_i \psi_{0i}(s) ,$$

where β_i are arbitrary constants.

We prove in passing that eigenfunctions $g(s)$ and $\psi(s)$ corresponding to distinct eigenvalues λ_1 and λ_2, respectively, of the homogeneous integral equation (3) and its transpose (15) are orthogonal. In fact, we have

$$g(s) = \lambda_1 \int K(s,t) g(t) \, dt \, , \qquad \psi(s) = \lambda_2 \int K(t,s) \psi(t) \, dt.$$

Multiplying both sides of the first equation by $\lambda_2 \psi(s)$ and those of the second equation by $\lambda_1 g(s)$, integrating, and then subtracting the resulting equations, we obtain

$$(\lambda_2 - \lambda_1) \int_a^b g(s) \psi(s) \, ds = 0 \, .$$

But $\lambda_1 \neq \lambda_2$, and the result follows.

We are now ready to discuss the solution of the inhomogeneous Fredholm integral equation (2.1.2) for the case $D(\lambda) = 0$. In fact, we can prove that the necessary and sufficient condition for this equation to have a solution for $\lambda = \lambda_0$, a root of $D(\lambda) = 0$, is that $f(s)$ be orthogonal to the r eigenfunctions ψ_{0i} of the transposed equation (15).

The necessary part of the proof follows from the fact that, if equation (2.1.2) for $\lambda = \lambda_0$ admits a certain solution $g(s)$, then

$$\int f(s) \psi_{0i}(s) \, ds = \int g(s) \psi_{0i}(s) \, ds$$
$$- \lambda_0 \int \psi_{0i}(s) \, ds \int K(s,t) g(t) \, dt$$
$$= \int g(s) \psi_{0i}(s) \, ds$$
$$- \lambda_0 \int g(t) \, dt \int K(s,t) \psi_{0i}(s) \, ds = 0 \, ,$$

because λ_0 and $\psi_{0i}(s)$ are eigenvalues and corresponding eigenfunctions of the transposed equation.

To prove the sufficiency of this condition, we again appeal to linear algebra. In fact, the corresponding condition of orthogonality for the linear-algebraic system assures us that the inhomogeneous system (8) reduces to only $n-r$ independent equations. This means that the rank of the matrix $(\mathbf{I} - \lambda \mathbf{A})$ is exactly $p = n - r$, and therefore the system (8) or (2.1.9) is soluble. Substituting this solution in (2.1.5), we have the solution to our integral equation.

Finally, the difference of any two solutions of (2.1.2) is a solution of the homogeneous equation (3). Hence, the most general solution of the inhomogeneous integral equation (2.1.2) has the form

$$g(s) = G(s) + \alpha_1 g_{01}(s) + \alpha_2 g_{02}(s) + \cdots + \alpha_r g_{0r}(s) , \qquad (16)$$

where $G(s)$ is a suitable linear combination of the functions $a_1(s)$, $a_2(s), \cdots, a_n(s)$.

We have thus proved the theorem that, if $\lambda = \lambda_0$ is a root of multiplicity $m \geqslant 1$ of the equation $D(\lambda) = 0$, then the inhomogeneous equation has a solution if and only if the given function $f(s)$ is orthogonal to all the eigenfunctions of the transposed equation.

The results of this section can be collected to establish the following theorem.

Fredholm Alternative Theorem.   Either the integral equation

$$g(s) = f(s) + \lambda \int K(s,t) g(t) \, dt \qquad (17)$$

with fixed λ possesses one and only one solution $g(s)$ for arbitrary $\mathscr{L}_2$-functions $f(s)$ and $K(s,t)$, in particular the solution $g = 0$ for $f = 0$; or the homogeneous equation

$$g(s) = \lambda \int K(s,t) g(t) \, dt \qquad (18)$$

possesses a finite number r of linearly independent solutions g_{0i}, $i = 1, 2, \cdots, r$. In the first case, the transposed inhomogeneous equation

$$\psi(s) = f(s) + \lambda \int K(t,s) \psi(t) \, dt \qquad (19)$$

also possesses a unique solution. In the second case, the transposed homogeneous equation

$$\psi(s) = \lambda \int K(t,s) \psi(t) \, dt \qquad (20)$$

also has r linearly independent solutions ψ_{0i}, $i = 1, 2, \cdots, r$; the inhomogeneous integral equation (7) has a solution if and only if the given function $f(s)$ satisfies the r conditions

$$(f, \psi_{0i}) = \int f(s) \psi_{0i}(s) \, ds = 0 , \qquad i = 1, 2, \cdots, r . \qquad (21)$$

In this case, the solution of (17) is determined only up to an additive linear combination $\sum_{i=1}^{r} c_i g_{0i}$.

The following examples illustrate the theorems of this section.

2.4. EXAMPLES

Example 1. Show that the integral equation

$$g(s) = f(s) + (1/\pi) \int_0^{2\pi} [\sin(s+t)] \, g(t) \, dt \qquad (1)$$

possesses no solution for $f(s) = s$, but that it possesses infinitely many solutions when $f(s) = 1$.

For this equation,

$$K(s, t) = \sin s \cos t + \cos s \sin t \,,$$

$$a_1(s) = \sin s \,, \qquad a_2(s) = \cos s \,, \qquad b_1(t) = \cos t \,, \qquad b_2(t) = \sin t \,.$$

Therefore,

$$a_{11} = \int_0^{2\pi} \sin t \cos t \, dt = 0 = a_{22} \,,$$

$$a_{12} = \int_0^{2\pi} \cos^2 t \, dt = \pi = a_{21} \,.$$

$$D(\lambda) = \begin{vmatrix} 1 & -\lambda\pi \\ -\lambda\pi & 1 \end{vmatrix} = 1 - \lambda^2 \pi^2 \,. \qquad (2)$$

The eigenvalues are $\lambda_1 = 1/\pi$, $\lambda_2 = -1/\pi$ and equation (1) contains $\lambda_1 = 1/\pi$. Therefore, we have to examine the eigenfunctions of the transposed equation (note that the kernel is symmetric)

$$g(s) = (1/\pi) \int_0^{2\pi} \sin(s+t) \, g(t) \, dt \,. \qquad (3)$$

The algebraic system corresponding to (3) is

$$c_1 - \lambda\pi c_2 = 0 \,, \qquad -\lambda\pi c_1 + c_2 = 0 \,,$$

which gives

$$c_1 = c_2 \quad \text{for} \quad \lambda_1 = 1/\pi \,; \qquad c_1 = -c_2 \quad \text{for} \quad \lambda_2 = -1/\pi \,.$$

Therefore, the eigenfunctions for $\lambda_1 = 1/\pi$ follow from the relation (2.1.5) and are given by

$$g(s) = c(\sin s + \cos s) . \tag{4}$$

Since

$$\int_0^{2\pi} (s \sin s + s \cos s) \, ds = -2\pi \neq 0 ,$$

while

$$\int_0^{2\pi} (\sin s + \cos s) \, ds = 0 ,$$

we have proved the result.

Example 2. Solve the integral equation

$$g(s) = f(s) + \lambda \int_0^1 (1 - 3st) g(t) \, dt . \tag{5}$$

The algebraic system (2.1.9) for this equation is

$$(1 - \lambda) c_1 + \tfrac{3}{2}\lambda c_2 = f_1 , \qquad -\tfrac{1}{2}\lambda c_1 + (1 + \lambda) c_2 = f_2 , \tag{6}$$

while

$$D(\lambda) = \begin{vmatrix} 1 - \lambda & \tfrac{3}{2}\lambda \\ -\tfrac{1}{2}\lambda & 1 + \lambda \end{vmatrix} = \tfrac{1}{4}(4 - \lambda^2) . \tag{7}$$

Therefore, the inhomogeneous equation (5) will have a unique solution if and only if $\lambda \neq \pm 2$. Then the homogeneous equation

$$g(s) = \lambda \int_0^1 (1 - 3st) g(t) \, dt \tag{8}$$

has only the trivial solution.

Let us now consider the case when λ is equal to one of the eigenvalues and examine the eigenfunctions of the transposed homogeneous equation

$$g(s) = \lambda \int_0^1 (1 - 3st) g(t) \, dt . \tag{9}$$

For $\lambda = +2$, the algebraic system (6) gives $c_1 = 3c_2$. Then, (2.1.5) gives the eigenfunction

$$g(s) = c(1 - s) , \tag{10}$$

where c is an arbitrary constant. Similarly, for $\lambda = -2$, the corresponding eigenfunction is

$$g(s) = c(1 - 3s) .\tag{11}$$

It follows from the above analysis that the integral equation

$$g(s) = f(s) + 2 \int_0^1 (1 - 3st) g(t) \, dt$$

will have a solution if $f(s)$ satisfies the condition

$$\int_0^1 (1 - s) f(s) \, ds = 0 ,$$

while the integral equation

$$g(s) = f(s) - 2 \int_0^1 (1 - 3st) g(t) \, dt$$

will have a solution if the following holds:

$$\int_0^1 (1 - 3s) f(s) \, ds = 0 .$$

2.5. AN APPROXIMATE METHOD

The method of this chapter is useful in finding approximate solutions of certain integral equations. We illustrate it by the following example:

$$g(s) = e^s - s - \int_0^1 s(e^{st} - 1) g(t) \, dt .\tag{1}$$

Let us approximate the kernel by the sum of the first three terms in its Taylor series:

$$K(t, s) = s(e^{st} - 1) \simeq s^2 t + \tfrac{1}{2} s^3 t^2 + \tfrac{1}{6} s^4 t^3 ,\tag{2}$$

that is, by a separable kernel. Then the integral equation takes the form

$$g(s) = e^s - s - \int_0^1 (s^2 t + \tfrac{1}{2} s^3 t^2 + \tfrac{1}{6} s^4 t^3) g(t) \, dt .\tag{3}$$

Since the kernel is separable, we require the solution in the form

$$g(s) = e^s - s + c_1 s^2 + c_2 s^3 + c_3 s^4 . \tag{4}$$

Following the method of this chapter, we find that the constants c_1, c_2, c_3 satisfy the following algebraic system:

$$(5/4) c_1 + (1/5) c_2 + (1/6) c_3 = -2/3 ,$$
$$(1/5) c_1 + (13/6) c_2 + (1/7) c_3 = (9/4) - e , \tag{5}$$
$$(1/6) c_1 + (1/7) c_2 + (49/8) c_3 = 2e - (29/5) ,$$

whose solution is

$$c_1 = -0.5010 , \qquad c_2 = -0.1671 , \qquad c_3 = -0.0422 . \tag{6}$$

Substituting these values in (4), we have the solution

$$g(s) = e^s - s - 0.5010 s^2 - 0.1671 s^3 - 0.0423 s^4 . \tag{7}$$

Now the exact solution of the integral equation (1) is

$$g(s) = 1 . \tag{8}$$

Using the approximate solution for $s = 0$, $s = 0.5$, and $s = 1.0$, the value of $g(s)$ from (7) is

$$g(0) = 1.0000 , \qquad g(0.5) = 1.0000 , \qquad g(1) = 1.0080 , \tag{9}$$

which agrees with (8) rather closely.

In Chapter 7 (see Section 7.7), we shall prove that an arbitrary $\mathscr{L}_2$-kernel can be approximated in norm by a separable kernel.

EXERCISES

1. Consider the equation

$$g(s) = f(s) + \lambda \int_0^1 K(s, t) g(t) \, dt$$

and show that for the kernels given in Table I the function $D(\lambda)$ has the given expression.

TABLE I

Case	Kernel	$D(\lambda)$
(i)	± 1	$1 \mp \lambda$
(ii)	st	$1 - (\lambda/3)$
(iii)	$s^2 + t^2$	$1 - (2\lambda/3) - (4\lambda^2/45)$

2. Solve the integral equation

$$g(s) = f(s) + \lambda \int_0^{2\pi} \cos(s+t) g(t) \, dt$$

and find the condition that $f(s)$ must satisfy in order that this equation has a solution when λ is an eigenvalue. Obtain the general solution if $f(s) = \sin s$, considering all possible cases.

3. Show that the integral equation

$$g(s) = \lambda \int_0^{\pi} (\sin s \sin 2t) g(t) \, dt$$

has no eigenvalues.

4. Solve the integral equation

$$g(s) = 1 + \lambda \int_{-\pi}^{\pi} e^{i\omega(s-t)} g(t) \, dt \, ,$$

considering separately all the exceptional cases.

5. In the integral equation

$$g(s) = s^2 + \int_0^1 (\sin st) g(t) \, dt \, ,$$

replace $\sin st$ by the first two terms of its power-series development

$$\sin st = st - \frac{(st)^3}{3!} + \cdots$$

and obtain an approximate solution.

METHOD OF SUCCESSIVE APPROXIMATIONS

3.1. ITERATIVE SCHEME

Ordinary first-order differential equations can be solved by the well-known Picard method of successive approximations. An iterative scheme based on the same principle is also available for linear integral equations of the second kind:

$$g(s) = f(s) + \lambda \int K(s,t) g(t) \, dt \, . \qquad (1)$$

In this chapter, we present this method. We assume the functions $f(s)$ and $K(s,t)$ to be $\mathscr{L}_2$-functions as defined in Chapter 1.

As a zero-order approximation to the desired function $g(s)$, the solution $g_0(s)$,

$$g_0(s) = f(s) \, , \qquad (2)$$

is taken. This is substituted into the right side of equation (1) to give the first-order approximation

$$g_1(s) = f(s) + \lambda \int K(s,t) g_0(t) \, dt \, . \qquad (3)$$

This function, when substituted into (1), yields the second approximation. This process is then repeated; the $(n+1)$th approximation is obtained by substituting the nth approximation in the right side of (1). There results the recurrence relation

$$g_{n+1}(s) = f(s) + \lambda \int K(s, t) g_n(t)\, dt . \tag{4}$$

If $g_n(s)$ tends uniformly to a limit as $n \to \infty$, then this limit is the required solution. To study such a limit, let us examine the iterative procedure (4) in detail. The first- and second-order approximations are

$$g_1(s) = f(s) + \lambda \int K(s, t) f(t)\, dt \tag{5}$$

and

$$g_2(s) = f(s) + \lambda \int K(s, t) f(t)\, dt$$
$$+ \lambda^2 \int K(s, t) \left[\int K(t, x) f(x)\, dx \right] dt . \tag{6}$$

This formula can be simplified by setting

$$K_2(s, t) = \int K(s, x) K(x, t)\, dx \tag{7}$$

and by changing the order of integration. The result is

$$g_2(s) = f(s) + \lambda \int K(s, t) f(t)\, dt + \lambda^2 \int K_2(s, t) f(t)\, dt . \tag{8}$$

Similarly,

$$g_3(s) = f(s) + \lambda \int K(s, t) f(t)\, dt$$
$$+ \lambda^2 \int K_2(s, t) f(t)\, dt + \lambda^3 \int K_3(s, t) f(t)\, dt , \tag{9}$$

where

$$K_3(s, t) = \int K(s, x) K_2(x, t)\, dx . \tag{10}$$

By continuing this process, and denoting

$$K_m(s, t) = \int K(s, x) K_{m-1}(x, t)\, dx , \tag{11}$$

we get the $(n+1)$th approximate solution of integral equation (1) as

$$g_n(s) = f(s) + \sum_{m=1}^{n} \lambda^m \int K_m(s, t) f(t)\, dt . \tag{12}$$

We call the expression $K_m(s, t)$ the mth iterate, where $K_1(s, t) = K(s, t)$. Passing to the limit as $n \to \infty$, we obtain the so-called Neumann series

$$g(s) = \lim_{n \to \infty} g_n(s) = f(s) + \sum_{m=1}^{\infty} \lambda^m \int K_m(s, t) f(t)\, dt . \tag{13}$$

It remains to determine conditions under which this convergence is achieved. For this purpose, we attend to the partial sum (12) and apply the Schwarz inequality (1.6.3) to the general term of this sum. This gives

$$\left| \int K_m(s,t) f(t) \, dt \right|^2 \leqslant \left(\int |K_m(s,t)|^2 \, dt \right) \int |f(t)|^2 \, dt . \tag{14}$$

Let D be the norm of f,

$$D^2 = \int |f(t)|^2 \, dt , \tag{15}$$

and let $C_m{}^2$ denote the upper bound of the integral

$$\int |K_m(s,t)|^2 \, dt .$$

Hence, the inequality (14) becomes

$$\left| \int K_m(s,t) f(t) \, dt \right|^2 \leqslant C_m{}^2 \, D^2 . \tag{16}$$

The next step is to connect the estimate $C_m{}^2$ with the estimate $C_1{}^2$. This is achieved by applying the Schwarz inequality to the relation (11):

$$|K_m(s,t)|^2 \leqslant \int |K_{m-1}(s,x)|^2 \, dx \int |K(x,t)|^2 \, dx ,$$

which, when integrated with respect to t, yields

$$\int |K_m(s,t)|^2 \, dt \leqslant B^2 \, C_{m-1}^2 , \tag{17}$$

where

$$B^2 = \int\!\!\int |K(x,t)|^2 \, dx \, dt . \tag{18}$$

The inequality (17) sets up the recurrence relation

$$C_m{}^2 \leqslant B^{2m-2} \, C_1{}^2 . \tag{19}$$

From (16) and (19), we have the inequality

$$\left| \int K_m(s,t) f(t) \, dt \right|^2 \leqslant C_1{}^2 \, D^2 \, B^{2m-2} . \tag{20}$$

Therefore, the general term of the partial sum (12) has a magnitude less than the quantity $DC_1 |\lambda|^m B^{m-1}$, and it follows that the infinite series (13) converges faster than the geometric series with common ratio $|\lambda| B$. Hence, if

$$|\lambda| B < 1 , \tag{21}$$

the uniform convergence of this series is assured.

It will now be proved that, for given λ, equation (1) has a unique solution. Suppose the contrary, and let $g_1(s)$ and $g_2(s)$ be two solutions of equation (1):

$$g_1(s) = f(s) + \lambda \int K(s,t) g_1(t) \, dt \, ,$$

$$g_2(s) = f(s) + \lambda \int K(s,t) g_2(t) \, dt \, .$$

By subtracting these equations and setting $g_1(s) - g_2(s) = \phi(s)$, there results the homogeneous integral equation

$$\phi(s) = \lambda \int K(s,t) \phi(t) \, dt \, .$$

Apply the Schwarz inequality to this equation and get

$$|\phi(s)|^2 \leqslant |\lambda|^2 \int |K(s,t)|^2 \, dt \int |\phi(t)|^2 \, dt \, ,$$

which, when integrated with respect to s, becomes

$$\int |\phi(s)|^2 \, ds \leqslant |\lambda|^2 \iint |K(s,t)|^2 \, ds \, dt \int |\phi(s)|^2 \, ds$$

or

$$(1 - |\lambda|^2 B^2) \int |\phi(s)|^2 \, ds \leqslant 0 \, . \tag{22}$$

In view of the inequality (21) and the nature of the function $\phi(s) = g_1(s) - g_2(s)$, we readily conclude that $\phi(s) \equiv 0$, i.e., $g_1(s) \equiv g_2(s)$.

What is the estimate of the error for neglecting terms after the nth term in the Neumann series (13)? This is found by writing this series as

$$g(s) = f(s) + \sum_{m=1}^{n} \lambda^m \int K_m(s,t) f(t) \, dt + R_n(s) \, . \tag{23}$$

Then, it follows from the above analysis that

$$|R_n| \leqslant DC_1 |\lambda|^{n+1} B^n / (1 - |\lambda| B) \, . \tag{24}$$

Finally, we can evaluate the resolvent kernel, as defined in the previous chapter, in terms of the iterated kernels $K_m(s, t)$. Indeed, by changing the order of integration and summation in the Neumann series (13), we obtain

$$g(s) = f(s) + \lambda \int \left[\sum_{m=1}^{\infty} \lambda^{m-1} K_m(s,t) \right] f(t) \, dt \, .$$

Comparing this with (2.3.7),

$$g(s) = f(s) + \lambda \int \Gamma(s,t;\lambda) f(t) \, dt \, , \tag{25}$$

we have

$$\Gamma(s, t; \lambda) = \sum_{m=1}^{\infty} \lambda^{m-1} K_m(s, t) . \tag{26}$$

From the above discussion, we can infer (see Section 3.5) that the series (26) is also convergent at least for $|\lambda| B < 1$. Hence, the resolvent kernel is an analytic function of λ, regular at least inside the circle $|\lambda| < B^{-1}$.

From the uniqueness of the solution of (1), we can prove that the resolvent kernel $\Gamma(s, t; \lambda)$ is unique. In fact, let equation (1) have, with $\lambda = \lambda_0$, two resolvent kernels $\Gamma_1(s, t; \lambda_0)$ and $\Gamma_2(s, t; \lambda_0)$. In view of the uniqueness of the solution of (1), an arbitrary function $f(s)$ satisfies the identity

$$f(s) + \lambda_0 \int \Gamma_1(s, t; \lambda_0) f(t) \, dt \equiv f(s) + \lambda_0 \int \Gamma_2(s, t; \lambda_0) f(t) \, dt . \tag{27}$$

Setting $\Psi(s, t; \lambda_0) = \Gamma_1(s, t; \lambda_0) - \Gamma_2(s, t; \lambda_0)$, we have, from (27),

$$\int \Psi(s, t; \lambda_0) f(t) \, dt \equiv 0 ,$$

for an arbitrary function $f(t)$. Let us take $f(t) = \Psi^*(s, t; \lambda)$, with fixed s. This implies that

$$\int |\Psi(s, t; \lambda_0)|^2 \, dt \equiv 0 ,$$

which means that $\Psi(s, t; \lambda_0) \equiv 0$, proving the uniqueness of the resolvent kernel.

The above analysis can be summed up in the following basic theorem.

Theorem. To each $\mathscr{L}_2$-kernel $K(s, t)$, there corresponds a unique resolvent kernel $\Gamma(s, t; \lambda)$ which is an analytic function of λ, regular at least inside the circle $|\lambda| < B^{-1}$, and represented by the power series (26). Furthermore, if $f(s)$ is also an $\mathscr{L}_2$-function, then the unique, $\mathscr{L}_2$-solution of the Fredholm equation (1) valid in the circle $|\lambda| < B^{-1}$ is given by the formula (25).

The method of successive approximations has many drawbacks. In addition to being a cumbersome process, the Neumann series, in general, cannot be summed in closed form. Furthermore, a solution of the integral equation (1) may exist even if $|\lambda| B > 1$, as evidenced in the previous chapter. In fact, we saw that the resolvent kernel is a quotient of two polynomials of nth degree in λ, and therefore the only possible singular points of $\Gamma(s, t; \lambda)$ are the roots of the denominator $D(\lambda) = 0$. But, for $|\lambda| B > 1$, the Neumann series does not converge and as such does not provide the desired solution. We shall have more to say about these ideas in the next chapter.

3.2. EXAMPLES

Example 1. Solve the integral equation

$$g(s) = f(s) + \lambda \int_0^1 e^{s-t} g(t)\, dt . \tag{1}$$

Following the method of the previous section, we have

$$K_1(s, t) = e^{s-t} ,$$

$$K_2(s, t) = \int_0^1 e^{s-x} e^{x-t}\, dx = e^{s-t} .$$

Proceeding in this way, we find that all the iterated kernels coincide with $K(s, t)$. Using (3.1.26), we obtain the resolvent kernel as

$$\Gamma(s, t; \lambda) = K(s, t)(1 + \lambda + \lambda^2 + \cdots) = e^{s-t}/(1 - \lambda) . \tag{2}$$

Although the series $(1 + \lambda + \lambda^2 + \cdots)$ converges only for $|\lambda| < 1$, the resolvent kernel is, in fact, an analytic function of λ, regular in the whole plane except at the point $\lambda = 1$, which is a simple pole of the kernel Γ. The solution $g(s)$ then follows from (3.1.25):

$$g(s) = f(s) - [\lambda/(\lambda - 1)] \int_0^1 e^{s-t} f(t)\, dt . \tag{3}$$

Example 2. Solve the Fredholm integral equation

$$g(s) = 1 + \lambda \int_0^1 (1 - 3st) g(t)\, dt \tag{4}$$

and evaluate the resolvent kernel.

Starting with $g_0(s) = 1$, we have

$$g_1(s) = 1 + \lambda \int_0^1 (1 - 3st)\, dt = 1 + \lambda(1 - \tfrac{3}{2}s)\,,$$

$$g_2(s) = 1 + \lambda \int_0^1 (1 - 3st)[1 + \lambda(1 - \tfrac{3}{2}t)]\, dt = 1 + \lambda(1 - \tfrac{3}{2}s) + \tfrac{1}{4}\lambda^2\,,$$

...

$$g(s) = 1 + \lambda(1 - \tfrac{3}{2}s) + \tfrac{1}{4}\lambda^2 + \tfrac{1}{4}\lambda^3(1 - \tfrac{3}{2}s) + \tfrac{1}{16}\lambda^4 + \tfrac{1}{16}\lambda^5(1 - \tfrac{3}{2}s) + \cdots,$$

or

$$g(s) = (1 + \tfrac{1}{4}\lambda^2 + \tfrac{1}{16}\lambda^4 + \cdots)[1 + \lambda(1 - \tfrac{3}{2}s)]\,. \tag{5}$$

The geometric series in (4) is convergent provided $|\lambda| < 2$. Then,

$$g(s) = [4 + 2\lambda(2 - 3s)]/(4 - \lambda^2)\,, \tag{6}$$

and precisely the same remarks apply to the region of the validity of this solution as given in Example 1.

To evaluate the resolvent kernel, we find the iterated kernels

$$K_1(s, t) = 1 - 3st\,,$$

$$K_2(s, t) = \int_0^1 (1 - 3sx)(1 - 3xt)\, dx = 1 - \tfrac{3}{2}(s + t) + 3st\,,$$

$$K_3(s, t) = \int_0^1 (1 - 3sx)[1 - \tfrac{3}{2}(x + t) - 3xt]\, dx$$

$$= \tfrac{1}{4}(1 - 3st) = \tfrac{1}{4}K_1(s, t)\,,$$

Similarly,

$$K_4(s, t) = \tfrac{1}{4}K_2(s, t)$$

and

$$K_n(s, t) = \tfrac{1}{4}K_{n-2}(s, t)\,.$$

Hence,

$$\Gamma(s, t; \lambda) = K_1 + \lambda K_2 + \lambda^2 K_3 + \cdots$$

$$= (1 + \tfrac{1}{4}\lambda^2 + \tfrac{1}{16}\lambda^4 + \cdots)K_1 + \lambda(1 + \tfrac{1}{4}\lambda^2 + \tfrac{1}{16}\lambda^4 + \cdots)K_2$$

$$= [(1 + \lambda) - \tfrac{3}{2}\lambda(s + t) - 3(1 - \lambda)st]/(1 - \tfrac{1}{4}\lambda^2)\,, \tag{7}$$

$$|\lambda| < 2\,.$$

Example 3.   Solve the integral equation

$$g(s) = 1 + \lambda \int_0^\pi [\sin(s+t)] g(t) \, dt \ . \tag{8}$$

Let us first evaluate the iterated kernels in this example:

$$K_1(s,t) = K(s,t) = \sin(s+t) \ ,$$

$$K_2(s,t) = \int_0^\pi [\sin(s+x)] \sin(x+t) \, dx$$

$$= \tfrac{1}{2}\pi [\sin s \sin t + \cos s \cos t] = \tfrac{1}{2}\pi \cos(s-t) \ ,$$

$$K_3(s,t) = \tfrac{1}{2}\pi \int_0^1 [\sin(s+x)] \cos(x-t) \, dx$$

$$= \tfrac{1}{2}\pi \int_0^1 (\sin s \cos x + \sin x \cos s)$$

$$\times (\cos x \cos t + \sin x \sin t) \, dx$$

$$= (\tfrac{1}{2}\pi)^2 [\sin s \cos t + \cos s \sin t] = (\tfrac{1}{2}\pi)^2 \sin(s+t) \ .$$

Proceeding in this manner, we obtain

$$K_4(s,t) = (\tfrac{1}{2}\pi)^3 \cos(s-t) \ ,$$
$$K_5(s,t) = (\tfrac{1}{2}\pi)^4 \sin(s+t) \ ,$$
$$K_6(s,t) = (\tfrac{1}{2}\pi)^5 \cos(s-t) \ , \quad \text{etc.}$$

Substituting these values in the formula (3.1.13) and integrating, there results the solution

$$g(s) = 2\lambda(\cos s)[1 + (\tfrac{1}{2}\pi)^2 \lambda^2 + (\tfrac{1}{2}\pi)^4 \lambda^4 + \cdots]$$
$$+ \lambda^2 \pi(\sin s)[1 + (\tfrac{1}{2}\pi)^2 \lambda^2 + (\tfrac{1}{2}\pi)^4 \lambda^4 + \cdots] \ , \tag{9}$$

or

$$g(s) = 1 + [(2\lambda \cos s + \lambda^2 \pi \sin s)/(1 - \tfrac{1}{4}\lambda^2 \pi^2)] \ . \tag{10}$$

Since

$$B^2 = \int_0^\pi \int_0^\pi \sin^2(s+t) \, ds \, dt = \tfrac{1}{2}\pi^2 \ ,$$

the interval of convergence of the series (9) lies between $-\sqrt{2}/\pi$ and $\sqrt{2}/\pi$.

Example 4. Prove that the mth iterated kernel $K_m(s, t)$ satisfies the following relation:

$$K_m(s, t) = \int K_r(s, x)\, K_{m-r}(x, t)\, dx \,, \tag{11}$$

where r is any positive integer less than n.

By successive applications of (3.1.11),

$$K_m(s, t) = \int \cdots \int K(s, x_1)\, K(x_1, x_2) \cdots K(x_{m-1}, t)\, dx_{m-1} \cdots dx_1 \,. \tag{12}$$

Thus, $K_m(s, t)$ is an $(m-1)$-fold integral. Similarly, $K_r(s, x)$ and $K_{m-r}(x, t)$ are $(r-1)$- and $(m-r-1)$-fold integrals. This means that

$$\int K_r(s, x)\, K_{m-r}(x, t)\, dx$$

is an $(m-1)$-fold integral, and the result follows.

One can take the $g_0(s)$ approximation different from $f(s)$, as we demonstrate by the following example.

Example 5. Solve the inhomogeneous Fredholm integral equation of the second kind,

$$g(s) = 2s + \lambda \int_0^1 (s+t)\, g(t)\, dt \,, \tag{13}$$

by the method of successive approximations to the third order.

For this equation, we take $g_0(s) = 1$. Then,

$$g(s) = 2s + \lambda \int_0^1 (s+t)\, dt = 2s + \lambda(s + \tfrac{1}{2}) \,,$$

$$g_2(s) = 2s + \lambda \int_0^1 (s+t)\, \{2t + \lambda[t + (1/2)]\}\, dt$$

$$= 2s + \lambda[s + (2/3)] + \lambda^2[s + (7/12)] \,,$$

$$g_3(s) = 2s + \lambda \int_0^1 (s+t)\, \{2t + \lambda[t + (2/3)] + \lambda^2[t + (7/12)]\}\, dt$$

$$= 2s + \lambda[s + (2/3)] + \lambda^2[(7/6)s + (2/3)] + \lambda^3[(13/12)s + (5/8)] \,. \tag{14}$$

From Example 2 of Section 2.2, we find the exact solution to be

$$g(s) = [12(2-\lambda)s + 8\lambda]/(12-2\lambda-\lambda^2) , \tag{15}$$

and the comparison of (14) and (15) is left to the reader.

3.3. VOLTERRA INTEGRAL EQUATION

The same iterative scheme is applicable to the Volterra integral equation of the second kind. In fact, the formulas corresponding to (3.1.13) and (3.1.25) are, respectively,

$$g(s) = f(s) + \sum_{m=1}^{\infty} \lambda^m \int_a^s K_m(s,t)f(t) \, dt , \tag{1}$$

$$g(s) = f(s) + \lambda \int_a^s \Gamma(s,t;\lambda)f(t) \, dt , \tag{2}$$

where the iterated kernel $K_m(s,t)$ satisfies the recurrence formula

$$K_m(s,t) = \int_t^s K(s,x) K_{m-1}(x,t) \, dx \tag{3}$$

with $K_1(s,t) = K(s,t)$, as before. The resolvent kernel $\Gamma(s,t;\lambda)$ is given by the same formula as (3.1.26), and it is an entire function of λ for any given (s,t) (see Exercise 8).

We shall illustrate it by the following examples.

3.4. EXAMPLES

Example 1.   Find the Neumann series for the solution of the integral equation

$$g(s) = (1+s) + \lambda \int_0^s (s-t)g(t) \, dt . \tag{1}$$

From the formula (3.3.3), we have

$$K_1(s,t) = (s-t) ,$$

$$K_2(s,t) = \int_t^s (s-x)(x-t)\,dx = \frac{(s-t)^3}{3!} ,$$

$$K_3(s,t) = \int_t^s \frac{(s-x)(x-t)^3}{3!}\,dx = \frac{(s-t)^5}{5!} ,$$

and so on. Thus,

$$g(s) = 1 + s + \lambda\left(\frac{s^2}{2!}+\frac{s^3}{3!}\right) + \lambda^2\left(\frac{s^4}{4!}+\frac{s^5}{5!}\right) + \cdots . \qquad (2)$$

For $\lambda = 1$, $g(s) = e^s$.

Example 2. Solve the integral equation

$$g(s) = f(s) + \lambda \int_0^s e^{s-t} g(t)\,dt \qquad (3)$$

and evaluate the resolvant kernel.

For this case,

$$K_1(s,t) = e^{s-t} ,$$

$$K_2(s,t) = \int_t^s e^{s-x} e^{x-t}\,dx = (s-t)e^{s-t} ,$$

$$K_3(s,t) = \int_t^s (x-t)e^{s-x} e^{x-t}\,dx = \frac{(s-t)^2}{2!} e^{s-t} ,$$

$$K_m(s,t) = \frac{(s-t)^{m-1}}{(m-1)!} e^{s-t} .$$

The resolvent kernel is

$$\Gamma(s,t;\lambda) = \begin{cases} e^{s-t} \displaystyle\sum_{m=1}^{\infty} \frac{\lambda^{m-1}(s-t)^{m-1}}{(m-1)!} = e^{(\lambda+1)(s-t)} , & t \leqslant s , \\ 0 , & t > s . \end{cases} \qquad (4)$$

Hence, the solution is

$$g(s) = f(s) + \lambda \int_0^s e^{(\lambda+1)(s-t)} f(t) \, dt \, . \tag{5}$$

Example 3.   Solve the Volterra equation

$$g(s) = 1 + \int_0^s st g(t) \, dt \, . \tag{6}$$

For this example, $K_1(s, t) = K(s, t) = st$,

$$K_2(s, t) = \int_t^s sx^2 t \, dx = (s^4 t - st^4)/3 \, ,$$

$$K_3(s, t) = \int_t^s [(sx)(x^4 t - xt^4)/3] \, dx = (s^7 t - 2s^4 t^4 + st^7)/18 \, ,$$

$$K_4(s, t) = \int_t^s [(sx)(x^7 t - 2x^4 t^4 + xt^7)/18] \, dx$$

$$= (s^{10} t - 3s^7 t + 3s^4 t^7 - st^{10})/162 \, ,$$

and so on. Thus.

$$g(s) = 1 + \frac{s^3}{3} + \frac{s^6}{2 \cdot 5} + \frac{s^9}{2 \cdot 5 \cdot 8} + \frac{s^{12}}{2 \cdot 5 \cdot 8 \cdot 11} + \cdots \, . \tag{7}$$

3.5. SOME RESULTS ABOUT THE RESOLVENT KERNEL

The series for the resolvent kernel $\Gamma(s, t; \lambda)$,

$$\Gamma(s, t; \lambda) = \sum_{m=1}^{\infty} \lambda^{m-1} K_m(s, t) \, , \tag{1}$$

can be proved to be absolutely and uniformly convergent for all values of s and t in the circle $|\lambda| < 1/B$. In addition to the assumptions of Section 3.1, we need the additional inequality

$$\int |K(s, t)|^2 \, ds < E^2 \, , \qquad E = \text{const} \, . \tag{2}$$

Recall that this is one of the conditions for the kernel K to be an $\mathscr{L}_2$-kernel. Applying the Schwarz inequality to the recurrence formula

$$K_m(s, t) = \int K_{m-1}(s, x) K(x, t) \, dx \qquad (3)$$

yields

$$|K_m(s, t)|^2 \leqslant \left(\int |K_{m-1}(s, x)|^2 \, dx \right) \int |K(x, t)|^2 \, dx \, ,$$

which, with the help of (3.1.19), becomes

$$|K_m(s, t)| \leqslant C_1 E B^{m-1} \, . \qquad (4)$$

Thus, the series (1) is dominated by the geometric series with the general term $C_1 E(\lambda^{m-1} B^{m-1})$, and that completes the proof.

Next, we prove that the resolvent kernel satisfies the integral equation

$$\Gamma(s, t; \lambda) = K(s, t) + \lambda \int K(s, x) \Gamma(x, t; \lambda) \, dx \, . \qquad (5)$$

This follows by replacing $K_m(s, t)$ in the series (1) by the integral relation (3). Then,

$$\Gamma(s, t; \lambda) = K_1(s, t) + \sum_{m=2}^{\infty} \lambda^{m-1} \int K_{m-1}(s, x) K(x, t) \, dx$$

$$= K(s, t) + \lambda \sum_{m=1}^{\infty} \lambda^{m-1} \int K_m(s, x) K(x, t) \, dx$$

$$= K(s, t) + \lambda \int [\sum_{m=1}^{\infty} \lambda^{m-1} K_m(s, x)] K(x, t) \, dx \, ,$$

and the integral equation (5) follows immediately. The change of order of integration and summation is legitimate in view of the uniform convergence of the series involved.

Another interesting result for the resolvent kernel is that it satisfies the integrodifferential equation

$$\partial \Gamma(s, t; \lambda) / \partial \lambda = \int \Gamma(s, x; \lambda) \Gamma(x, t; \lambda) \, dx \, . \qquad (6)$$

In fact,

$$\int \Gamma(s, x; \lambda) \Gamma(x, t; \lambda) dx = \int \sum_{m=1}^{\infty} \lambda^{m-1} K_m(s, x) \sum_{n=1}^{\infty} \lambda^{n-1} K_n(x, t) \, dx \, .$$

On account of the absolute and uniform convergence of the series (1), we can multiply the series under the integral sign and integrate it term by term. Therefore,

$$\int \Gamma(s, x; \lambda) \Gamma(x, t; \lambda) \, dx = \sum_{m=1}^{\infty} \sum_{n=1}^{\infty} \lambda^{m+n-2} K_{m+n}(s, t) \, . \qquad (7)$$

Now, set $m + n = p$ and change the order of summation; there results the relation

$$\sum_{m=1}^{\infty} \sum_{n=1}^{\infty} \lambda^{m+n-2} K_{m+n}(s, t) = \sum_{p=2}^{\infty} \sum_{n=1}^{p-1} \lambda^{p-2} K_p(s, t)$$

$$= \sum_{p=2}^{\infty} (p-1) \lambda^{p-2} K_p(s, t) = \frac{\partial \Gamma(s, t; \lambda)}{\partial \lambda} \, . \qquad (8)$$

Combining (7) and (8), we have the result (6).

EXERCISES

1. Solve the following Fredholm integral equations by the method of successive approximations:

$$\text{(i)} \quad g(s) = e^s - \tfrac{1}{2}e + \tfrac{1}{2} + \tfrac{1}{2} \int_0^1 g(t) \, dt \, .$$

$$\text{(ii)} \quad g(s) = (\sin s) - \tfrac{1}{4}s + \tfrac{1}{4} \int_0^{\pi/2} stg(t) \, dt \, .$$

2. Consider the integral equation

$$g(s) = 1 + \lambda \int_0^1 stg(t) \, dt \, .$$

(a) Make use of the relation $|\lambda| < B^{-1}$ to show that the iterative procedure is valid for $|\lambda| < 3$.

(b) Show that the iterative procedure leads formally to the solution

$$g(s) = 1 + s[(\lambda/2) + (\lambda^2/6) + (\lambda^3/18) + \cdots] \, .$$

(c) Use the method of the previous chapter to obtain the exact solution

$$g(s) = 1 + [3\lambda s/2(3 - \lambda)] \, , \qquad \lambda \neq 3 \, .$$

3. Solve the integral equation

$$g(s) = 1 + \lambda \int_0^1 (s+t) g(t) \, dt \, ,$$

by the method of successive approximations and show that the estimate afforded by the relation $|\lambda| < B^{-1}$ is conservative in this case.

4. Find the resolvent kernel associated with the following kernels: (i) $|s-t|$, in the interval $(0, 1)$; (ii) $\exp - |s-t|$, in the interval $(0, 1)$; (iii) $\cos(s+t)$, in the interval $(0, 2\pi)$.

5. Solve the following Volterra integral equations by the method of this chapter:

$$\text{(i)} \quad g(s) = 1 + \int_0^s (s-t) g(t) \, dt \, ,$$

$$\text{(ii)} \quad g(s) = 29 + 6s + \int_0^s (6s - 6t + 5) g(t) \, dt \, .$$

6. Find an approximate solution of the integral equation

$$g(s) = (\sinh s) + \int_0^s e^{t-s} g(t) \, dt \, ,$$

by the method of iteration.

7. Obtain the radius of convergence of the Neumann series when the functions $f(s)$ and the kernel $K(s, t)$ are continuous in the interval (a, b).

8. Prove that the resolvent kernel for a Volterra integral equation of the second kind is an entire function of λ for any given (s, t).

CLASSICAL FREDHOLM THEORY

4.1. THE METHOD OF SOLUTION OF FREDHOLM

In the previous chapter, we have derived the solution of the Fredholm integral equation

$$g(s) = f(s) + \lambda \int K(s,t) g(t) \, dt \tag{1}$$

as a uniformly convergent power series in the parameter λ for $|\lambda|$ suitably small. Fredholm gave the solution of equation (1) in general form for all values of the parameter λ. His results are contained in three theorems which bear his name. We have already studied them in Chapter 2 for the special case when the kernel is separable. In this chapter, we shall study equation (1) when the function $f(s)$ and the kernel $K(s,t)$ are any integrable functions. Furthermore, the present method enables us to get explicit formulas for the solution in terms of certain determinants.

The method used by Fredholm consists in viewing the integral equation (1) as the limiting case of a system of linear algebraic equations. This theory applies to two- or higher-dimensional integrals, although we shall confine our discussion to only one-dimensional integrals in the interval (a, b). Let us divide the interval (a, b) into n equal parts,

$$s_1 = t_1 = a, \quad s_2 = t_2 = a+h, \quad \ldots, \quad s_n = t_n = a+(n-1)h,$$

where $h = (b-a)/n$. Thereby, we have the approximate formula

$$\int K(s,t) g(t) \, dt \simeq h \sum_{j=1}^{n} K(s, s_j) g(s_j) \, . \tag{2}$$

Equation (1) then takes the form

$$g(s) \simeq f(s) + \lambda h \sum_{j=1}^{n} K(s, s_j) g(s_j) \, , \tag{3}$$

which must hold for all values of s in the interval (a, b). In particular, this equation is satisfied at the n points of division s_i, $i = 1, ..., n$. This leads to the system of equations

$$g(s_i) = f(s_i) + \lambda h \sum_{j=1}^{n} K(s_i, s_j) g(s_j) \, , \qquad i = 1, ..., n \, . \tag{4}$$

Writing

$$f(s_i) = f_i \, , \qquad g(s_i) = g_i \, , \qquad K(s_i, s_j) = K_{ij} \, , \tag{5}$$

equation (4) yields an approximation for the integral equation (1) in terms of the system of n linear equations

$$g_i - \lambda h \sum_{j=1}^{n} K_{ij} g_j = f_i \, , \qquad i = 1, ..., n \, , \tag{6}$$

in n unknown quantities $g_1, ..., g_n$. The values of g_i obtained by solving this algebraic system are approximate solutions of the integral equation (1) at the points $s_1, s_2, ..., s_n$. We can plot these solutions g_i as ordinates and by interpolation draw a curve $g(s)$ which we may expect to be an approximation to the actual solution. With the help of this algebraic system, we can also determine approximations for the eigenvalues of the kernel.

The resolvent determinant of the algebraic system (6) is

$$D_n(\lambda) = \begin{vmatrix} 1 - \lambda h K_{11} & -\lambda h K_{12} & \cdots & -\lambda h K_{1n} \\ -\lambda h K_{21} & 1 - \lambda h K_{22} & \cdots & -\lambda h K_{2n} \\ \vdots & & & \\ -\lambda h K_{n1} & -\lambda h K_{n2} & \cdots & 1 - \lambda h K_{nn} \end{vmatrix} . \tag{7}$$

The approximate eigenvalues are obtained by setting this determinant equal to zero. We illustrate it by the following example.

Example.

$$g(s) - \lambda \int_0^{\pi} \sin(s+t) g(t) \, dt = 0 \, .$$

By taking $n = 3$, we have $h = \pi/3$ and therefore

$$s_1 = t_1 = 0 \, , \qquad s_2 = t_2 = \pi/3 \, , \qquad s_3 = t_3 = 2\pi/3 \, ,$$

and the values of K_{ij} are readily calculated to give

$$(K_{ij}) = \begin{vmatrix} 0 & 0.866 & 0.866 \\ 0.866 & 0.866 & 0 \\ 0.866 & 0 & -0.866 \end{vmatrix} .$$

The homogeneous system corresponding to (6) will have a non-trivial solution if the determinant

$$D_n(\lambda) = \begin{vmatrix} 1 & -0.907\lambda & -0.907\lambda \\ -0.907\lambda & (1-0.907\lambda) & 0 \\ -0.907\lambda & 0 & (1+0.907\lambda) \end{vmatrix} = 0 \, ,$$

or when $1 - 3(0.0907)^2 \lambda^2 = 0$. The roots of this equation are $\lambda = \pm 0.6365$. This gives a rather close agreement with the exact values (see Example 3, Section 3.2), which are $\pm \sqrt{2}/\pi = \pm 0.6366$.

In general, the practical applications of this method are limited because one has to take a rather large n to get a reasonable approximation.

4.2. FREDHOLM'S FIRST THEOREM

The solutions $g_1, g_2, \ldots, g_n$ of the system of equations (4.1.6) are obtained as ratios of certain determinants, with the determinant $D_n(\lambda)$ given by (4.1.7) as the denominator provided it does not vanish. Let us expand the determinant (4.1.7) in powers of the quantity $(-\lambda h)$. The constant term is obviously equal to unity. The term containing $(-\lambda h)$ in the first power is the sum of all the determinants containing only one column $-\lambda h K_{\mu\nu}$, $\mu = 1, \ldots, n$. Taking the contribution from all the columns $\nu = 1, \ldots, n$, we find that the total contribution is $-\lambda h \sum_{\nu=1}^{n} K_{\nu\nu}$.

The factor containing the factor $(-\lambda h)$ to the second power is the sum of all the determinants containing two columns with that factor. This results in the determinants of the form

$$(-\lambda h)^2 \begin{vmatrix} K_{pp} & K_{pq} \\ K_{qp} & K_{qq} \end{vmatrix},$$

where (p,q) is an arbitrary pair of integers taken from the sequence $1, \ldots, n$, with $p < q$. In the same way, it follows that the term containing the factor $(-\lambda h)^3$ is the sum of the determinants of the form

$$(-\lambda h)^3 \begin{vmatrix} K_{pp} & K_{pq} & K_{pr} \\ K_{qp} & K_{qq} & K_{qr} \\ K_{rp} & K_{rq} & K_{rr} \end{vmatrix},$$

where (p, q, r) is an arbitrary triplet of integers selected from the sequence $1, \ldots, n$, with $p < q < r$.

The remaining terms are obtained in a similar manner. Therefore, we conclude that the required expansion of $D_n(\lambda)$ is

$$D_n(\lambda) = 1 - \lambda h \sum_{\nu=1}^{n} K_{\nu\nu} + \frac{(-\lambda h)^2}{2!} \sum_{p,q=1}^{n} \begin{vmatrix} K_{pp} & K_{pq} \\ K_{qp} & K_{qq} \end{vmatrix}$$

$$+ \frac{(-\lambda h)^3}{3!} \sum_{p,q,r=1}^{n} \begin{vmatrix} K_{pp} & K_{pq} & K_{pr} \\ K_{qp} & K_{qq} & K_{qr} \\ K_{rp} & K_{rq} & K_{rr} \end{vmatrix} + \cdots$$

$$+ \frac{(-\lambda h)^n}{n!} \sum_{p_1,p_2,\ldots,p_n=1}^{n} \begin{vmatrix} K_{p_1 p_1} & K_{p_1 p_2} & \cdots & K_{p_1 p_n} \\ K_{p_2 p_1} & K_{p_2 p_2} & \cdots & K_{p_2 p_n} \\ \vdots & & & \\ K_{p_n p_1} & K_{p_n p_2} & \cdots & K_{p_n p_n} \end{vmatrix}, \qquad (1)$$

where we now stipulate that the sums are taken over all permutations of pairs (p,q), triplets (p,q,r), etc. This convention explains the reason for dividing each term of the above series by the corresponding number of permutations.

The analysis is simplified by introducing the following symbol for the determinant formed by the values of the kernel at all points (s_i, t_j)

$$\begin{vmatrix} K(s_1, t_1) & K(s_1, t_2) & \cdots & K(s_1, t_n) \\ K(s_2, t_1) & K(s_2, t_2) & \cdots & K(s_2, t_n) \\ \vdots & & & \\ K(s_n, t_1) & K(s_n, t_2) & \cdots & K(s_n, t_n) \end{vmatrix} = K\begin{pmatrix} s_1, s_2, \ldots, s_n \\ t_1, t_2, \ldots, t_n \end{pmatrix}, \quad (2)$$

the so-called Fredholm determinant. We observe that, if any pair of arguments in the upper or lower sequence is transposed, the value of the determinant changes sign because the transposition of two arguments in the upper sequence corresponds to the transposition of two rows of the determinant and the transposition of two arguments in the lower sequence corresponds to the transposition of two columns.

In this notation, the series (1) takes the form

$$D_n(\lambda) = 1 - \lambda h \sum_{p=1}^{n} K(s_p, s_p) + \frac{(-\lambda h)^2}{2!} \sum_{p,q=1}^{n} K\begin{pmatrix} s_p, s_q \\ s_p, s_q \end{pmatrix}$$

$$+ \frac{(-\lambda h)^3}{3!} \sum_{p,q,r=1}^{n} K\begin{pmatrix} s_p, s_q, s_r \\ s_p, s_q, s_r \end{pmatrix} + \cdots . \quad (3)$$

If we now let n tend to infinity, then h will tend to zero, and each term of the sum (3) tends to some single, double, triple integral, etc. There results Fredholm's first series:

$$D(\lambda) = 1 - \lambda \int K(s, s)\, ds + \frac{\lambda^2}{2!} \int K\begin{pmatrix} s_1, s_2 \\ s_1, s_2 \end{pmatrix} ds_1\, ds_2$$

$$- \frac{\lambda^3}{3!} \iiint K\begin{pmatrix} s_1, s_2, s_3 \\ s_1, s_2, s_3 \end{pmatrix} ds_1\, ds_2\, ds_3 + \cdots . \quad (4)$$

Hilbert gave a rigorous proof of the fact that the sequence $D_n(\lambda) \to D(\lambda)$ in the limit, while the convergence of the series (4) for all values of λ was proved by Fredholm on the basis that the kernel $K(s, t)$ is a bounded and integrable function.[1] Thus, $D(\lambda)$ is an entire function of the complex variable λ.

We are now ready to solve the Fredholm equation (4.1.1) and express

---

[1] For proof, see Lovitt [10].

the solutions in the form of a quotient of two power series in the parameter λ, where the Fredholm function $D(\lambda)$ is to be the divisor. In this connection, recall the relations (2.3.6) and (2.3.7). Indeed, we seek solutions of the form

$$g(s) = f(s) + \lambda \int \Gamma(s, t; \lambda) f(t) \, dt \,, \tag{5}$$

and expect the resolvent kernel $\Gamma(s, t; \lambda)$ to be the quotient

$$\Gamma(s, t; \lambda) = D(s, t; \lambda) / D(\lambda) \,, \tag{6}$$

where $D(s, t; \lambda)$, still to be determined, is the sum of certain functional series.

Now, we have proved in Section 3.5 that the resolvent $\Gamma(s, t; \lambda)$ itself satisfies a Fredholm integral equation of the second kind (3.5.5):

$$\Gamma(s, t; \lambda) = K(s, t) + \lambda \int K(s, x) \Gamma(x, t; \lambda) \, dx \,. \tag{7}$$

From (6) and (7), it follows that

$$D(s, t; \lambda) = K(s, t) D(\lambda) + \lambda \int K(s, x) D(x, t; \lambda) \, dx \,. \tag{8}$$

The form of the series (4) for $D(\lambda)$ suggests that we seek the solution of equation (8) in the form of a power series in the parameter λ:

$$D(s, t; \lambda) = C_0(s, t) + \sum_{p=1}^{\infty} \frac{(-\lambda)^p}{p!} C_p(s, t) \,. \tag{9}$$

For this purpose, write the numerical series (4) as

$$D(\lambda) = 1 + \sum_{p=1}^{\infty} \frac{(-\lambda)^p}{p!} c_p \,, \tag{10}$$

where

$$c_p = \int \cdots \int K \begin{pmatrix} s_1, s_2, \ldots, s_p \\ s_1, s_2, \ldots, s_p \end{pmatrix} ds_1 \cdots ds_p \,. \tag{11}$$

The next step is to substitute the series for $D(s, t; \lambda)$ and $D(\lambda)$ from (9) and (10) in (8) and compare the coefficients of equal powers of λ. The following relations result:

$$C_0(s, t) = K(s, t) \,, \tag{12}$$

$$C_p(s, t) = c_p K(s, t) - p \int K(s, x) C_{p-1}(x, t) \, dx \,. \tag{13}$$

Our contention is that we can write the function $C_p(s, t)$ in terms of the Fredholm determinant (2) in the following way:

$$C_p(s, t) = \int \cdots \int K\begin{pmatrix} s, x_1, x_2, \ldots, x_p \\ t, x_1, x_2, \ldots, x_p \end{pmatrix} dx_1 \cdots dx_p . \tag{14}$$

In fact, for $p = 1$, the relation (13) becomes

$$
\begin{aligned}
C_1(s, t) &= c_1 K(s\ t) - \int K(s, x) C_0(x, t)\, dx \\
&= K(s, t) \int K(x, x)\, dx - \int K(s, x) K(x, t)\, dx \\
&= \int K\begin{pmatrix} s & x \\ t & x \end{pmatrix} dx ,
\end{aligned}
\tag{15}
$$

where we have used (11) and (12).

To prove that (14) holds for general p, we expand the determinant under the integral sign in the relation:

$$
K\begin{pmatrix} s, x_1, \ldots, x_p \\ t, x_1, \ldots, x_p \end{pmatrix} = \begin{vmatrix} K(s, t) & K(s, x_1) & \cdots & K(s, x_p) \\ K(x_1, t) & K(x_1, x_1) & \cdots & K(x_1, x_p) \\ \vdots & & & \\ K(x_p, t) & K(x_p, x_1) & \cdots & K(x_p, x_p) \end{vmatrix},
$$

with respect to the elements of the given row, transposing in turn the first column one place to the right, integrating both sides, and using the definition of c_p as in (11); the required result then follows by induction.

From (9), (12), and (14) we derive Fredholm's second series:

$$D(s, t; \lambda) = K(s, t) + \sum_{p=1}^{\infty} \frac{(-\lambda)^p}{p!} \int \cdots \int K\begin{pmatrix} s, x_1, \ldots, x_p \\ t, x_1, \ldots, x_p \end{pmatrix} dx_1 \cdots dx_p . \tag{16}$$

This series also converges for all values of the parameter λ. It is interesting to observe the similarity between the series (4) and (16).

Having found both terms of the quotient (6), we have established the existence of a solution to the integral equation (4.1.1) for a bounded and integrable kernel $K(s, t)$, provided, of course, that $D(\lambda) \neq 0$. Since both terms of this quotient are entire functions of the parameter λ, it follows that the resolvent kernel $\Gamma(s, t; \lambda)$ is a meromorphic function of λ, i.e., an analytic function whose singularities may only be the poles, which in the present case are zeros of the divisor $D(\lambda)$.

Next, we prove that the solution in the form obtained by Fredholm is unique and is given by

$$g(s) = f(s) + \lambda \int \Gamma(s, t; \lambda) f(t)\, dt . \tag{17}$$

In this connection, we first observe that the integral equation (7) satisfied by $\Gamma(s,t;\lambda)$ is valid for all values of λ for which $D(\lambda) \neq 0$. Indeed, (7) is known to hold for $|\lambda| < B^{-1}$ from the analysis of Chapter 3, and since both sides of this equation are now proved to be meromorphic, the above contention follows. To prove the uniqueness of the solution, let us suppose that $g(s)$ is a solution of the equation (4.1.1) in the case $D(\lambda) \neq 0$. Multiply both sides of (4.1.1) by $\Gamma(s,t;\lambda)$, integrate, and get

$$\int \Gamma(s,x;\lambda)g(x)\,dx = \int \Gamma(s,x;\lambda)f(x)\,dx$$
$$+ \lambda \int \left[\int \Gamma(s,x;\lambda)K(x,t)\,dx \right] g(t)\,dt. \quad (18)$$

Substituting from (7) into left side of (18), this becomes

$$\int K(s,t)g(t)\,dt = \int \Gamma(s,x;\lambda)f(x)\,dx, \quad (19)$$

which, when joined by (4.1.1), yields

$$g(s) = f(s) + \lambda \int \Gamma(s,t;\lambda)f(t)\,dt, \quad (20)$$

but this form is unique.

In particular, the solution of the homogeneous equation

$$g(s) = \lambda \int K(s,t)g(t)\,dt \quad (21)$$

is identically zero.

The above analysis leads to the following theorem.

Fredholm's First Theorem. The inhomogeneous Fredholm equation

$$g(s) = f(s) + \lambda \int K(s,t)g(t)\,dt, \quad (22)$$

where the functions $f(s)$ and $g(t)$ are integrable, has a unique solution

$$g(s) = f(s) + \lambda \int \Gamma(s,t;\lambda)f(t)\,dt, \quad (23)$$

where the resolvent kernel $\Gamma(s,t;\lambda)$,

$$\Gamma(s,t;\lambda) = D(s,t;\lambda)/D(\lambda), \quad (24)$$

with $D(\lambda) \neq 0$, is a meromorphic function of the complex variable λ, being the ratio of two entire functions defined by the series

$$D(s,t;\lambda) = K(s,t) + \sum_{p=1}^{\infty} \frac{(-\lambda)^p}{p!} \int \cdots \int K\binom{s,x_1,\ldots,x_p}{t,x_1,\ldots,x_p} dx_1 \cdots dx_p, \quad (25)$$

and

$$D(\lambda) = 1 + \sum_{p=1}^{\infty} \frac{(-\lambda)^p}{p!} \int \cdots \int K\binom{x_1, \ldots, x_p}{x_1, \ldots, x_p} dx_1 \cdots dx_p , \qquad (26)$$

both of which converge for all values of λ. In particular, the solution of the homogeneous equation

$$g(s) = \lambda \int K(s, t) g(t) \, dt \qquad (27)$$

is identically zero.

4.3. EXAMPLES

Example 1.   Evaluate the resolvent for the integral equation

$$g(s) = f(s) + \lambda \int_0^1 (s + t) g(t) \, dt . \qquad (1)$$

The solution to this example is obtained by writing

$$\Gamma(s, t; \lambda) = \left[\sum_{p=0}^{\infty} \frac{(-\lambda)^p}{p!} C_p(s, t) \right] \bigg/ \sum_{p=0}^{\infty} \frac{(-\lambda)^p}{p!} c_p , \qquad (2)$$

where C_p and c_p are defined by the relations (4.2.11) and (4.2.13):

$$c_0 = 1 , \qquad C_0(s, t) = K(s, t) = (s + t) . \qquad (3)$$

$$c_p = \int C_{p-1}(s, s) \, ds , \qquad (4)$$

$$C_p = c_p K(s, t) - p \int_0^1 K(s, x) c_{p-1}(x, t) \, dx . \qquad (5)$$

Thus,

$$c_1 = \int_0^1 2s \, ds = 1 ,$$

$$C_1(s, t) = (s + t) - \int_0^1 (s + x)(x + t) \, dx = \tfrac{1}{2}(s + t) - st - \tfrac{1}{3} ,$$

$$c_2 = \int_0^1 (s - s^2 - \tfrac{1}{3}) \, ds = -\tfrac{1}{6} ,$$

$$C_2(s,t) = -\tfrac{1}{6}(s+t) - 2 \int_0^1 (s+x)[\tfrac{1}{2}(x+t) - xt - \tfrac{1}{3})] \, dx = 0 \; .$$

Since $C_2(x,t)$ vanishes, it follows from (5) that the subsequent coefficients C_k and c_k also vanish. Therefore,

$$\Gamma(s,t;\lambda) = \frac{(s+t) - [\tfrac{1}{2}(s+t) - st - \tfrac{1}{3}]\lambda}{1 - \lambda - (\lambda^2/12)} \; , \tag{6}$$

which agrees with result (2.2.8) found by a different method.

Example 2. Solve the integral equation

$$g(s) = s + \lambda \int_0^1 [st + (st)^{\frac{1}{2}}] g(t) \, dt \; . \tag{7}$$

In this case,

$$c_0 = 1 \; , \qquad C_0(s,t) = st + (st)^{\frac{1}{2}} \; ,$$

$$c_1 = \int_0^1 (s^2 + s) \, ds = \tfrac{5}{6} \; ,$$

$$C_1(s,t) = \tfrac{5}{6}[st + (st)^{\frac{1}{2}}] - \int_0^1 [sx + (sx)^{\frac{1}{2}}][xt + (xt)^{\frac{1}{2}}] \, dt$$

$$= \tfrac{1}{2}st + \tfrac{1}{3}(st)^{\frac{1}{2}} - \tfrac{2}{5}(st^{\frac{1}{2}} + ts^{\frac{1}{2}}) \; ,$$

$$c_2 = \int_0^1 (\tfrac{1}{2}s^2 + \tfrac{1}{3}s - \tfrac{4}{5}s^{\frac{3}{2}}) \, ds = 1/75 \; ,$$

$$C_2(s,t) = 0 \; ,$$

and therefore all the subsequent coefficients vanish. The value of the resolvent is

$$\Gamma(s,t;\lambda) = \frac{st + (st)^{\frac{1}{2}} - \{\tfrac{1}{2}st + \tfrac{1}{3}(st)^{\frac{1}{2}} - \tfrac{2}{5}(st^{\frac{1}{2}} + s^{\frac{1}{2}}t)\}\lambda}{1 - \tfrac{5}{6}\lambda + (1/150)\lambda^2} \; . \tag{8}$$

The solution $g(s)$ then follows by using the relation (4.2.23),

$$g(s) = \frac{150s + \lambda(60\sqrt{s} - 75s) + 21\lambda^2 s}{\lambda^2 - 125\lambda + 150} \; . \tag{9}$$

Example 3. As a final example, let us solve the integral equation

$$g(s) = 1 + \lambda \int_0^\pi [\sin(s+t)]\, g(t)\, dt\,, \tag{10}$$

which we have already solved in Section 3.2 by a Neumann series.

For the present kernel,

$$c_0 = 1\,, \qquad C_0(s,t) = \sin(s+t)\,, \qquad c_1 = \int_0^\pi \sin 2s\, ds = 0\,.$$

$$C_1(s,t) = 0 - \int_0^\pi \sin(s+x)\sin(x+t)\, dx = -\tfrac{1}{2}\pi \cos(s-t)\,,$$

$$c_2 = -\int_0^\pi \tfrac{1}{2}\pi \cos 0\, ds = -\tfrac{1}{2}\pi^2\,,$$

$$C_2(s,t) = -\tfrac{1}{2}\pi^2 \sin(s+t) + 2 \int_0^\pi \tfrac{1}{2}\pi \sin(s+x)\cos(x-t)\, dx = 0\,.$$

Hence, the resolvent is

$$\Gamma(s,t;\lambda) = \frac{\sin(s+t) + \tfrac{1}{2}\pi\lambda \cos(s-t)}{1 - \tfrac{1}{4}\pi^2 \lambda^2}\,,$$

and the solution is

$$g(s) = 1 + \frac{2\lambda(\cos s) + \lambda^2 \pi(\sin s)}{1 - \tfrac{1}{4}\pi^2 \lambda^2}\,, \tag{11}$$

which agrees with the solution (3.2.10).

4.4. FREDHOLM'S SECOND THEOREM

Fredholm's first theorem does not hold when λ is a root of the equation $D(\lambda) = 0$. We have found in Chapter 2 that, for a separable kernel, the homogeneous equation

$$g(s) = \lambda \int K(s,t)\, g(t)\, dt \tag{1}$$

has nontrivial solutions. It might be expected that same holds when the kernel is an arbitrary integrable function and we shall then have a spectrum of eigenvalues and corresponding eigenfunctions. The second theorem of Fredholm is devoted to the study of this problem.

We first prove that every zero of $D(\lambda)$ is a pole of the resolvent kernel (4.2.24); the order of this pole is at most equal to the order of the zero of $D(\lambda)$. In fact, differentiate the Fredholm's first series (4.2.26) and interchange the indices of the variables of integration to get

$$D'(\lambda) = -\int D(s,s;\lambda)\,ds \ . \tag{2}$$

From this relation, it follows that, if λ_0 is a zero of order k of $D(\lambda)$, then it is a zero of order $k-1$ of $D'(\lambda)$ and consequently λ_0 may be a zero of order at most $k-1$ of the entire function $D(s,t;\lambda)$. Thus, λ_0 is the pole of the quotient (4.2.24) of order at most k. In particular, if λ_0 is a simple zero of $D(\lambda)$, then $D(\lambda_0) = 0$, $D'(\lambda_0) \neq 0$, and λ_0 is a simple pole of the resolvent kernel. Moreover, it follows from (2) that $D(s,t;\lambda) \neq 0$. For this particular case, we observe from equation (4.1.8) that, if $D(\lambda) = 0$ and $D(s,t;\lambda) \neq 0$, then $D(s,t;\lambda)$, as a function of s, is a solution of the homogeneous equation (1). So is $\alpha D(s,t;\lambda)$, where α is an arbitrary constant.

Let us now consider the general case when λ is a zero of an arbitrary multiplicity m, that is, when

$$D(\lambda_0) = 0\ , \quad \ldots\ , \quad D^{(r)}(\lambda_0) = 0\ , \quad D^{(m)}(\lambda_0) \neq 0\ , \tag{3}$$

where the superscript r stands for the differential of order r, $r = 1$, ..., $m-1$. For this case, the analysis is simplified if one defines a determinant known as the Fredholm minor:

$$
\begin{aligned}
D_n &\begin{pmatrix} s_1, s_2, \ldots, s_n \\ t_1, t_2, \ldots, t_n \end{pmatrix} \lambda \\
&= K\begin{pmatrix} s_1, s_2, \ldots, s_n \\ t_1, t_2, \ldots, t_n \end{pmatrix} + \sum_{p=1}^{\infty} \frac{(-\lambda)^p}{p!} \\
&\quad \times \int \cdots \int K\begin{pmatrix} s_1, \ldots, s_n, x_1, \ldots, x_p \\ t_1, \ldots, t_n, x_1, \ldots, x_p \end{pmatrix} dx_1\, dx_2 \cdots dx_p\ ,
\end{aligned} \tag{4}
$$

where $\{s_i\}$ and $\{t_i\}$, $i = 1, 2, \ldots, n$, are two sequences of arbitrary variables. Just as do the Fredholm series (4.2.25) and (4.2.26), the series (4) also converges for all values of λ and consequently is an entire function of λ. Furthermore, by differentiating the series (4.2.26) n times and comparing it with the series (4), there follows the relation

$$\frac{d^n D(\lambda)}{d\lambda^n} = (-1)^n \int \cdots \int D_n \left(\begin{matrix} s_1, \ldots, s_n \\ s_1, \ldots, s_n \end{matrix} \middle| \lambda \right) ds_1 \cdots ds_n \ . \tag{5}$$

From this relation, we conclude that, if λ_0 is a zero of multiplicity m of the function $D(\lambda)$, then the following holds for the Fredholm minor of order m for that value of λ_0:

$$D_m \left(\begin{matrix} s_1, s_2, \ldots, s_m \\ t_1, t_2, \ldots, t_m \end{matrix} \middle| \lambda_0 \right) \neq 0 \ .$$

Of course, there might exist minors of order lower than m which also do not identically vanish (compare the discussion in Section 2.3).

Let us find the relation among the minors that corresponds to the resolvent formula (4.2.7). Expansion of the determinant under the integral sign in (4),

$$\begin{vmatrix} K(s_1,t_1) & K(s_1,t_2) & \cdots & K(s_1,t_n) & K(s_1,x_1) & \cdots & K(s_1,x_p) \\ K(s_2,t_1) & K(s_2,t_2) & \cdots & K(s_2,t_n) & K(s_2,x_1) & \cdots & K(s_2,x_p) \\ \vdots & & & & & & \\ K(s_n,t_1) & K(s_n,t_2) & \cdots & K(s_n,t_n) & K(s_n,x_1) & \cdots & K(s_n,x_p) \\ K(x_1,t_1) & K(x_1,t_2) & \cdots & K(x_1,t_n) & K(x_1,x_1) & \cdots & K(x_1,x_p) \\ \vdots & & & & & & \\ K(x_p,t_1) & K(x_p,t_2) & \cdots & K(x_p,t_n) & K(x_p,x_1) & \cdots & K(x_p,x_p) \end{vmatrix} \tag{6}$$

by elements of the first row and integrating p times with respect to $x_1, x_2, \ldots, x_p$ for $p \geqslant 1$, we have

$$\int \cdots \int K \left(\begin{matrix} s_1, \ldots, s_n, x_1, \ldots, x_p \\ t_1, \ldots, t_n, x_1, \ldots, x_p \end{matrix} \right) dx_1 \cdots dx_p$$

$$= \sum_{h=1}^{n} (-1)^{h+1} K(s_1, t_h)$$

$$\times \int \cdots \int K \left(\begin{matrix} s_2, \ldots, s_h, \ldots, & s_n, x_1, \ldots, x_p \\ t_1, \ldots, t_{h-1}, t_{h+1}, \ldots, t_n, x_1, \ldots, x_p \end{matrix} \right) dx_1 \, dx_2 \cdots dx_p$$

$$+ \sum_{h=1}^{p} (-1)^{h+n-1}$$

$$\times \int \cdots \int K(s_1, x_h) \, K \left(\begin{matrix} s_2, \ldots, & s_n, x_1, & x_2, & \ldots, & x_h, & \ldots, & x_p \\ t_1, \ldots, t_{n-1}, t_n, x_1, \ldots, x_{h-1}, x_{h+1}, \ldots, x_p \end{matrix} \right)$$

$$\times dx_1 \cdots dx_p \ . \tag{7}$$

Note that the symbols for the determinant K on the right side of (7) do not contain the variables s_1 in the upper sequence and the variables t_h or x_h in the lower sequence. Furthermore, it follows by transposing the variable s_h in the upper sequence to the first place by means of $h+n-2$ transpositions that all the components of the second sum on the right side are equal. Therefore, we can write (7) as

$$\int \cdots \int K\begin{pmatrix} s_1, \ldots, s_n, x_1, \ldots, x_p \\ t_1, \ldots, t_n, x_1, \ldots, x_p \end{pmatrix}$$

$$= \sum_{h=1}^{n} (-1)^{h+1} K(s_1, t_h)$$

$$\times \int \cdots \int K\begin{pmatrix} s_2, \ldots, & s_n, x_1, \ldots, x_p \\ t_1, \ldots, t_{h-1}, t_{h+1}, \ldots, t_n, x_1, \ldots, x_p \end{pmatrix} dx_1 \cdots dx_p$$

$$- p \int K(s_1, x) \left[\int \cdots \int K\begin{pmatrix} x, s_2, \ldots, s_n, x_1, \ldots, x_{p-1} \\ t_1, t_2, \ldots, t_n, x_1, \ldots, x_{p-1} \end{pmatrix} \right.$$

$$\left. \times dx_1 \cdots dx_{p-1} \right] dx \tag{8}$$

where we have omitted the subscript h from x. Substituting (8) in (7), we find that Fredholm minor satisfies the integral equation

$$D_n \begin{pmatrix} s_1, \ldots, s_n \\ t_1, \ldots, t_n \end{pmatrix} \lambda = \sum_{h=1}^{n} (-1)^{h+1} K(s_1, t_h) D_{n-1} \begin{pmatrix} s_2, \ldots & , s_n \\ t_1, \ldots, t_{h-1}, t_{h+1}, t_n \end{pmatrix}$$

$$+ \lambda \int K(s_1, x) D_n \begin{pmatrix} x, x_2, \ldots, s_n \\ t_1, t_2, \ldots, t_n \end{pmatrix} \lambda dx . \tag{9}$$

Expansion by the elements of any other row leads to a similar identity, with x placed at the corresponding place. If we expand the determinant (6) with respect to the first column and proceed as above, we get the integral equation

$$D_n \begin{pmatrix} s_1, \ldots, s_n \\ t_1, \ldots, t_n \end{pmatrix} \lambda = \sum_{h=1}^{n} (-1)^{h+1} K(s_h, t_1) D_{n-1} \begin{pmatrix} s_1, \ldots, s_{h-1}, s_{h+1}, \ldots, s_n \\ t_2, \ldots & , t_n \end{pmatrix}$$

$$+ \lambda \int K(x, t_1) D_n \begin{pmatrix} s_1, \ldots & , s_n \\ x, t_2, \ldots, t_n \end{pmatrix} dx , \tag{10}$$

and a similar result would follow if we were to expand by any other column. The formulas (9) and (10) will play the role of the Fredholm series of the previous section.

Note that the relations (9) and (10) hold for all values of λ. With the help of (9), we can find the solution of the homogeneous equation (1) for the special case when $\lambda = \lambda_0$ is an eigenvalue. To this end, let us suppose that $\lambda = \lambda_0$ is a zero of multiplicity m of the function $D(\lambda)$. Then, as remarked earlier, the minor D_m does not identically vanish and even the minors $D_1, D_2, ..., D_{m-1}$ may not identically vanish. Let D_r be the first minor in the sequence $D_1, D_2, ..., D_{m-1}$ that does not vanish identically. The number r lies between 1 and m and is the index of the eigenvalue λ_0 as defined in Section 2.3. Moreover, this means that $D_{r-1} = 0$. But then the integral equation (9) implies that

$$g_1(s) = D_r\left(\begin{matrix} s, s_2, ..., s_r \\ t_1, ..., \quad t_r \end{matrix}\bigg| \lambda_0\right) \tag{11}$$

is a solution of the homogeneous equation (1). Substituting s at different points of the upper sequence in the minor D_r, we obtain r nontrivial solutions $g_i(s)$, $i = 1, ..., r$, of the homogeneous equation. These solutions are usually written as

$$\Phi_i(s) = \frac{D_r\left(\begin{matrix} s_1, ..., s_{i-1}, s, s_{i+1}, ..., s_r \\ t_1, ... \quad\quad\quad\quad\quad , t_r \end{matrix}\bigg| \lambda_0\right)}{D_r\left(\begin{matrix} s_1, ..., s_{i-1}, s_i, s_{i+1}, ..., s_r \\ t_1, ... \quad\quad\quad\quad\quad , t_r \end{matrix}\bigg| \lambda_0\right)}, \qquad i = 1, 2, ..., r. \tag{12}$$

Observe that we have already established that the denominator is not zero.

The solutions Φ_i as given by (12) are linearly independent for the following reason. In the determinant (6) above, if we put two of the arguments s_i equal, this amounts to putting two rows equal, and consequently the determinant vanishes. Thus, in (12), we see that $\Phi_k(s_i) = 0$ for $i \neq k$, whereas $\Phi_k(s_k) = 1$. Now, if there exists a relation $\sum_k C_k \Phi_k \equiv 0$, we may put $s = s_i$, and it follows that $C_i \equiv 0$; and this proves the linear independence of these solutions. This system of solutions Φ_i is called the fundamental system of the eigenfunctions of λ_0 and any linear combination of these functions gives a solution of (1).

Conversely, we can show that any solution of equation (1) must be a linear combination of $\Phi_1(s), \Phi_2(s), \ldots, \Phi_\nu(s)$. We need to define a kernel $H(s, t; \lambda)$ which corresponds to the resolvent kernel $\Gamma(s, t; \lambda)$ of the previous section

$$H(s, t; \lambda) = D_{r+1}\left(\begin{matrix} s, s_1, \ldots, s_r \\ t, t_1, \ldots, t_r \end{matrix} \middle| \lambda_0 \right) \middle/ D_r\left(\begin{matrix} s_1, \ldots, s_r \\ t_1, \ldots, t_r \end{matrix} \middle| \lambda_0 \right). \tag{13}$$

In (10), take n to be equal to r, and add extra arguments s and t to obtain

$$D_{r+1}\left(\begin{matrix} s, s_1, \ldots, s_r \\ t, t_1, \ldots, t_r \end{matrix} \middle| \lambda_0 \right) = K(s, t) D_r\left(\begin{matrix} s_1, \ldots, s_r \\ t_1, \ldots, t_r \end{matrix} \middle| \lambda_0 \right)$$

$$+ \sum_{h=1}^{r} (-1)^h K(s_h, y) D_r\left(\begin{matrix} s, s_1, \ldots, s_{h-1}, s_{h+1}, \ldots, s_r \\ t_1, t_2, \ldots \qquad\qquad\quad, t_r \end{matrix} \middle| \lambda_0 \right)$$

$$+ \lambda_0 \int K(x, t) D_{r+1}\left(\begin{matrix} s, s_1, \ldots, s_r \\ x, t_1, \ldots, t_r \end{matrix} \middle| \lambda_0 \right) dx. \tag{14}$$

In every minor D_r in the above equation, we transpose the variable s from the first place to the place between the variables s_{h-1} and s_{h+1} and divide both sides by the constant

$$D_r\left(\begin{matrix} s_1, \ldots, s_r \\ t_1, \ldots, t_r \end{matrix} \middle| \lambda_0 \right) \neq 0 ,$$

to obtain

$$H(s, t; \lambda) - K(s, t) - \lambda_0 \int H(s, x; \lambda) K(x, t) \, dx$$

$$= - \sum_{h=1}^{r} K(s_h, t) \Phi_h(s) . \tag{15}$$

If $g(s)$ is any solution to (1), we multiply (15) by $g(t)$ and integrate with respect to t,

$$\int g(t) H(s, t; \lambda) \, dt - \frac{g(s)}{\lambda_0} - \int g(x) \Gamma(s, x; \lambda) \, dx$$

$$= - \sum_{h=1}^{r} \frac{g(s_h)}{\lambda_0} \Phi_h(s) , \tag{16}$$

where we have used (1) in all terms but the first; we have also taken $\lambda_0 \int K(s_h, t) g(t) \, dt = g(s_h)$. Cancelling the equal terms, we have

$$g(s) = \sum_{h=1}^{r} g(s_h) \Phi_h(s) . \tag{17}$$

This proves our assertion. Thus we have established the following result.

Fredholm's Second Theorem. If λ_0 is a zero of multiplicity m of the function $D(\lambda)$, then the homogeneous equation

$$g(s) = \lambda_0 \int K(s,t) g(t) \, dt \tag{18}$$

possesses at least one, and at most m, linearly independent solutions

$$g_i(s) = D_r \left(\begin{matrix} s_1, \ldots, s_{i-1}, s, s_{i+1}, \ldots, s_r \\ t_1, \ldots \qquad\qquad\qquad , t_r \end{matrix} \middle| \lambda_0 \right) ,$$
$$i = 1, \ldots, r ; \qquad 1 \leqslant r \leqslant m \tag{19}$$

not identically zero. Any other solution of this equation is a linear combination of these solutions.

4.5. FREDHOLM'S THIRD THEOREM

In the analysis of Fredholm's first theorem, it has been shown that the inhomogeneous equation

$$g(s) = f(s) + \lambda \int K(s,t) g(t) \, dt \tag{1}$$

possesses a unique solution provided $D(\lambda) \neq 0$. Fredholm's second theorem is concerned with the study of the homogeneous equation

$$g(s) = \lambda \int K(s,t) g(t) \, dt ,$$

when $D(\lambda) = 0$. In this section, we investigate the possibility of (1) having a solution when $D(\lambda) = 0$. The analysis of this section is not much different from the corresponding analysis for separable kernels as given in Section 2.3. In fact, the only difference is that we shall now give an explicit formula for the solution. Qualitatively, the discussion is the same.

Recall that the transpose (or adjoint) of equation (1) is (under the same assumption as in Section 2.3)

$$\psi(s) = f(s) + \lambda \int_a^b K(t,s)\psi(t)\,dt \ . \tag{2}$$

It is clear that Fredholm's first series $D(\lambda)$ as given by (4.1.26) is the same for the transposed equation, while the second series is $D(t,s;\lambda)$ as obtained from (4.1.25) by interchanging the roles of s and t. This means that the kernels of equation (1) and its transpose (2) have the same eigenvalues. Furthermore, the resolvent kernel for (2) is

$$\Gamma(t,s;\lambda) = D(t,s;\lambda)/D(\lambda) \ , \tag{3}$$

and therefore the solution of (2) is

$$\psi(s) = f(s) + \lambda \int [D(t,s;\lambda)/D(\lambda)]f(t)\,dt \ , \tag{4}$$

provided λ is not an eigenvalue.

It is also clear that not only has the transposed kernel the same eigenvalues as the original kernel, but also the index r of each of the eigenvalues is equal. Moreover, corresponding to equation (4.4.12), the eigenfunctions of the transposed equation for an eigenvalue for λ_0 are given as

$$\Psi_i(t) = \frac{D_r\begin{pmatrix} s_1, \ldots & , s_r \\ t_1, \ldots, t_{i-1}, t, t_{i+1}, \ldots, t_r \end{pmatrix} \lambda_0}{D_r\begin{pmatrix} s_1, \ldots & , s_r \\ t_1, \ldots, t_{i-1}, t_i, t_{i+1}, \ldots, t_r \end{pmatrix} \lambda_0} \ , \tag{5}$$

where the values $(s_1, \ldots, s_r)$ and $(t_1, \ldots, t_r)$ are so chosen that the denominator does not vanish. Substituting r in different places in the lower sequence of this formula, we obtain a linearly independent system of r eigenfunctions. Also recall that each Φ_i is orthogonal to each Ψ_i with different eigenvalues.

If a solution $g(s)$ of (1) exists, then multiply (1) by each member $\Psi_k(s)$ of the above-mentioned system of functions and integrate to obtain

$$\int f(s)\Psi_k(s)\,ds = \frac{\int g(s)\Psi_k(s)\,ds - \lambda \iint K(s,t)g(t)\Psi_k(s)\,ds\,dt}{\int g(s)\,ds\,[\Psi_k(s) - \lambda \int K(t,s)\Psi_k(t)\,dt] = 0} \ , \tag{6}$$

where the term in the bracket vanishes because $\Psi_k(s)$ is an eigenfunction of the transposed equation. From (6), we see that a necessary condition for (1) to have a solution is that the inhomogeneous term $f(s)$ be orthogonal to each solution of the transposed homogeneous equation.

Conversely, we shall show that the condition (6) of orthogonality is sufficient for the existence of a solution. Indeed, we shall present an explicit solution in that case. With this purpose, we again appeal to the resolvent function $H(s, t; \lambda)$ as defined by (4.4.13) under the assumption that $D_r \neq 0$ and that r is the index of the eigenvalue λ_0.

Our contention is that if the orthogonality condition is satisfied, then the function

$$g_0(s) = f(s) + \lambda_0 \int H(s, t; \lambda) f(t) \, dt \tag{7}$$

is a solution. Indeed, substitute this value for $g(s)$ in (1), obtaining

$$f(s) + \lambda_0 \int H(s, t; \lambda) f(t) \, dt = f(s) + \lambda_0 \int K(s, t)$$
$$\times [f(t) + \lambda_0 \int H(t, x; \lambda) f(x) \, dx] \, dt$$

or

$$\int f(t) \, dt \, [H(s, t; \lambda) - K(s, t) - \lambda_0 \int K(s, x) H(x, t; \lambda) \, dx] = 0 . \tag{8}$$

Now, just as we obtained equation (4.4.15), we can obtain its "transpose,"

$$H(s, t; \lambda) - K(s, t) - \lambda_0 \int K(s, x) H(x, t; \lambda) \, dx$$
$$= - \sum_{h=1}^{r} K(s, t_h) \Psi_h(t) . \tag{9}$$

Substituting this in (8) and using the orthogonality condition, we have an identity, and thereby the assertion is proved.

The difference of any two solutions of (1) is a solution of the homogeneous equation. Hence, the most general solution of (1) is

$$g(s) = f(s) + \lambda_0 \int H(s, t; \lambda) f(t) \, dt + \sum_{h=1}^{r} C_h \Phi_h(s) . \tag{10}$$

The above analysis leads to the following theorem.

Fredholm's Third Theorem.   For an inhomogeneous equation

$$g(s) = f(s) + \lambda_0 \int K(s, t) g(t) \, dt , \tag{11}$$

to possess a solution in the case $D(\lambda_0) = 0$, it is necessary and sufficient that the given function $f(s)$ be orthogonal to all the eigenfunctions

$\Psi_i(s)$, $i = 1, 2, ..., \nu$, of the transposed homogeneous equation corresponding to the eigenvalue λ_0. The general solution has the form

$$g(s) = f(s) + \lambda_0 \int \left[D_{r+1}\begin{pmatrix} s, s_1, s_2, ..., s_r \\ t, t_1, ... \quad , t_r \end{pmatrix} \lambda_0 \right] \Bigg/ D_r\begin{pmatrix} s_1, s_2, ..., s_r \\ t_1, ... \quad , t_r \end{pmatrix} \lambda_0 \Bigg)$$

$$\times f(t) \, dt + \sum_{h=1}^{r} C_h \Phi_h(s) . \tag{12}$$

EXERCISES

1. Use the method of this chapter and find the resolvent kernels for the following three integral equations:

$$\text{(i)} \quad g(s) = f(s) + \lambda \int_0^1 |s-t| \, g(t) \, dt ,$$

$$\text{(ii)} \quad g(s) = f(s) + \lambda \int_0^1 \left(\exp - |s-t| \right) g(t) \, dt ,$$

$$\text{(iii)} \quad g(s) = f(s) + \lambda \int_0^{2\pi} \left[\cos(s+t) \right] g(t) \, dt .$$

2. Solve the following homogeneous equations

$$\text{(i)} \quad g(s) = \tfrac{1}{2} \int_0^{\pi} \left[\sin(s+t) \right] g(t) \, dt ,$$

$$\text{(ii)} \quad g(s) = \left[1/(e^2 - 1) \right] \int_0^1 2 e^s e^t g(t) \, dt ,$$

by using the explicit formulas of this chapter.

3. Show by the present method that the resolvent kernel for the integral equation with kernel $K(s, t) = 1 - 3st$, in the interval $(0, 1)$, is

$$\Gamma(s, t; \lambda) = [4/(4 - \lambda^2)] [1 + \lambda - \tfrac{3}{2}(s+t) - 3(1-\lambda)st] , \qquad \lambda \neq \pm 2 .$$

4. Show that not every one of the Fredholm's minors as defined by (4.4.4) is identically zero.

Hint: Use (4.4.5).

CHAPTER 5

APPLICATIONS TO ORDINARY DIFFERENTIAL EQUATIONS

The theories of ordinary and partial differential equations are fruitful sources of integral equations. In the quest for the representation formula for the solution of a linear differential equation in such a manner so as to include the boundary condition or initial condition explicitly, one is always led to an integral equation. Once a boundary value or an initial value problem has been formulated in terms of an integral equation, it becomes possible to solve this problem easily. In this chapter, we shall consider only ordinary differential equations. The next chapter is devoted to partial differential equations.

5.1. INITIAL VALUE PROBLEMS

There is a fundamental relationship between Volterra integral equations and ordinary differential equations with prescribed initial values. We begin our discussion by studying the simple initial value problem

$$y'' + A(s)y' + B(s)y = F(s), \tag{1}$$

$$y(a) = q_0, \qquad y'(a) = q_1, \tag{2}$$

where a prime implies differentiation with respect to s, and the functions A, B, and F are defined and continuous in the closed interval $a \leqslant s \leqslant b$.

The result of integrating the differential equation (1) from a to s and using the initial values (2) is

$$y'(s) - q_1 = -A(s)y(s) - \int_a^s [B(s_1) - A'(s_1)] y(s_1) \, ds_1$$

$$+ \int_a^s F(s_1) \, ds_1 + A(a)q_0 .$$

Similarly, a second integration yields

$$y(s) - q_0 = -\int_a^s A(s_1)y(s_1) \, ds_1 - \int_a^s \int_a^{s_2} [B(s_1) - A'(s_1)] y(s_1) \, ds_1 \, ds_2$$

$$+ \int_a^s \int_a^{s_2} F(s_1) \, ds_1 \, ds_2 + [A(a)q_0 + q_1](s-a) . \tag{3}$$

With the help of the identity (see Appendix, Section A.1)

$$\int_a^s \int_a^{s_2} F(s_1) \, ds_1 \, ds_2 = \int_a^s (s-t) F(t) \, dt , \tag{4}$$

the two double integrals in (3) can be converted to single integrals. Hence, the relation (3) takes the form

$$y(s) = q_0 + [A(a)q_0 + q_1](s-a) + \int_a^s (s-t) F(t) \, dt$$

$$- \int_a^s \{A(t) + (s-t)[B(t) - A'(t)]\} y(t) \, dt . \tag{5}$$

Now, set

$$K(s,t) = -\{A(t) + (s-t)[B(t) - A'(t)]\} \tag{6}$$

and

$$f(s) = \int_a^s (s-t) F(t) \, dt + [A(a)q_0 + q_1](s-a) + q_0 . \tag{7}$$

From relations (5)–(7), we have the Volterra integral equation of the second kind:

$$y(s) = f(s) + \int_a^s K(s, t)\, y(t)\, dt \ . \tag{8}$$

Conversely, any solution $g(s)$ of the integral equation (8) is, as can be verified by two differentiations, a solution of the initial value problem (1)–(2).

Note that the crucial step is the use of the identity (4). Since we have proved the corresponding identity for an arbitrary integer n in the Appendix, Section A.1, it follows that the above process of converting an initial value problem to a Volterra integral equation is applicable to a linear ordinary differential equation of order n when there are n prescribed initial conditions. An alternative approach is somewhat simpler for proving the above-mentioned equivalence for a general differential equation. Indeed, let us consider the linear differential equation of order n:

$$\frac{d^n y}{ds^n} + A_1(s)\frac{d^{n-1} y}{ds^{n-1}} + \cdots + A_{n-1}(s)\frac{dy}{ds} + A_n(s)y = F(s) , \tag{9}$$

with the initial conditions

$$y(a) = q_0 , \qquad y'(a) = q_1, \qquad \ldots, \qquad y^{(n-1)}(a) = q_{n-1} , \tag{10}$$

where the functions $A_1, A_2, \ldots, A_n$ and F are defined and continuous in $a \leqslant s \leqslant b$.

The reduction of the initial value problem (9)–(10) to the Volterra integral equation is accomplished by introducing an unknown function $g(s)$:

$$d^n y / ds^n = g(s) \ . \tag{11}$$

From (10) and (11), it follows that

$$\frac{d^{n-1}(y)}{ds^{n-1}} = \int_a^s g(t)\, dt + q_{n-1} ,$$

$$\frac{d^{n-2} y}{ds^{n-2}} = \int_a^s (s-t)g(t)\, dt + (s-a)q_{n-1} + q_{n-2} ,$$

$$\tag{12}$$

(continued)

...

$$\frac{dy}{ds} = \int_a^s \frac{(s-t)^{n-2}}{(n-2)!} g(t) \, dt + \frac{(s-a)^{n-2}}{(n-2)!} q_{n-1} + \frac{(s-a)^{n-3}}{(n-3)!} q_{n-2}$$

$$+ \cdots + (s-a)q_2 + q_1 \, ,$$

$$y = \int_a^s \frac{(s-t)^{n-1}}{(n-1)!} g(t) \, dt + \frac{(s-a)^{n-1}}{(n-1)!} q_{n-1} + \frac{(s-a)^{n-2}}{(n-2)!} q_{n-2}$$

$$+ \cdots + (s-a)q_1 + q_0 \, .$$

(12)

Now, if we multiply relations (11) and (12) by 1, $A_1(s)$, $A_2(s)$, etc. and add, we find that the initial value problem defined by (9)–(10) is reduced to the Volterra integral equation of the second kind

$$g(s) = f(s) + \int_a^s K(s, t) g(t) \, dt \, , \tag{13}$$

where

$$K(s, t) = -\sum_{k=1}^n A_k(s) \frac{(s-t)^{k-1}}{(k-1)!} \tag{14}$$

and

$$f(s) = F(s) - q_{n-1} A_1(s) - [(s-a)q_{n-1} + q_{n-2}] A_2(s)$$

$$- \cdots - \{[(s-a)^{n-1}/(n-1)!]q_{n-1} + \cdots + (s-a)q_1 + q_0\}$$

$$\times A_n(s) \, . \tag{15}$$

Conversely, if we solve the integral equation (13) and substitute the value obtained for $g(s)$ in the last equation of the system (12), we derive the (unique) solution of the initial value problem (9)–(10).

5.2. BOUNDARY VALUE PROBLEMS

Just as initial value problems in ordinary differential equations lead to Volterra-type integral equations, boundary value problems in ordinary differential equations lead to Fredholm-type integral equations. Let us illustrate this equivalence by the problem

$$y''(s) + A(s)y' + B(s)y = F(s) , \tag{1}$$

$$y(a) = y_0 , \qquad y(b) = y_1 . \tag{2}$$

When we integrate equation (1) from a to s and use the boundary condition $y(a) = y_0$, we get

$$y'(s) = C + \int_a^s F(s)\, ds - A(s)y(s) + A(a)y_0$$

$$+ \int_a^s [A'(s) - B(s)]y(s)\, ds ,$$

where C is a constant of integration.

A second integration similarly yields

$$y(s) - y_0 = [C + A(a)y_0](s-a) + \int_a^s \int_a^{s_2} F(s_1)\, ds_1\, ds_2$$

$$- \int_a^s A(s_1)y(s_1)\, ds_1 + \int_a^s \int_a^{s_2} [A'(s_1) - B(s_1)]y(s_1)\, ds_1\, ds_2 . \tag{3}$$

Using the identity (5.1.4), the relation (3) becomes

$$y(s) - y_0 = [C + A(a)y_0](s-a) + \int_a^s (s-t)F(t)\, dt$$

$$- \int_a^s \{A(t) - (s-t)[A'(t) - B(t)]\}y(t)\, dt . \tag{4}$$

The constant C can be evaluated by setting $s = b$ in (4) and using the second boundary condition $y(b) = y_1$:

$$y_1 - y_0 = [C + A(a)y_0](b-a) + \int (b-t)F(t)\, dt$$

$$- \int \{A(t) - (b-t)[A'(t) - B(t)]\}y(t)\, dt ,$$

or

$$C + A(a)y_0 = [1/(b-a)] \{(y_1 - y_0) - \int (b-t)F(t)\, dt$$

$$+ \int \{A(t) - (b-t)[A'(t) - B(t)]\} y(t)\, dt\} . \quad (5)$$

From (4) and (5), we have the relation

$$y(s) = y_0 + \int_a^s (s-t)F(t)\, dt + [(s-a)/(b-a)]$$

$$\times [(y_1 - y_0) - \int (b-t)F(t)\, dt]$$

$$- \int_a^s \{A(t) - (s-t)[A'(t) - B(t)]\} y(t)\, dt$$

$$+ \int [(s-a)/(b-a)] \{A(t) - (b-t)[A'(t) - B(t)]\} y(t)\, dt . \quad (6)$$

Equation (6) can be written as the Fredholm integral equation

$$y(s) = f(s) + \int K(s,t)\, y(t)\, dt , \quad (7)$$

provided we set

$$f(s) = y_0 + \int_a^s (s-t)F(t)\, dt$$

$$+ [(s-a)/(b-a)][(y_1 - y_0) - \int (b-t)F(t)\, dt] \quad (8)$$

and

$$K(s,t) = \begin{cases} [(s-a)/(b-a)]\{A(t) - (b-t)[A'(t) - B(t)]\} , \\ \qquad\qquad\qquad\qquad s < t , \\ A(t)\{[(s-a)/(b-a)] - 1\} - [A'(t) - B(t)] \\ \qquad \times [(t-a)(b-s)/(b-a)] , \quad s > t . \end{cases} \quad (9)$$

For the special case when A and B are constants, $a = 0$, $b = 1$, and $y(0) = y(1) = 0$, the above kernel simplifies to

$$K(s,t) = \begin{cases} Bs(1-t) + As , & s < t , \\ Bt(1-s) + As - A , & s > t . \end{cases} \quad (10)$$

Note that the kernel (10) is asymmetric and discontinuous at $t = s$, unless $A \equiv 0$. We shall elaborate on this point in Section 5.4.

5.3. EXAMPLES

Example 1.   Reduce the initial value problem

$$y''(s) + \lambda y(s) = F(s) , \tag{1}$$

$$y(0) = 1 , \qquad y'(0) = 0 , \tag{2}$$

to a Volterra integral equation.

Comparing (1) and (2) with the notation of Section 5.1, we have $A(s) = 0$, $B(s) = \lambda$. Therefore, the relations (5.1.6)–(5.1.8) become

$$K(s,t) = \lambda(t-s) ,$$

$$f(s) = 1 + \int_0^s (s-t) F(t) \, dt , \tag{3}$$

and

$$y(s) = 1 + \int_0^s (s-t) F(t) \, dt + \lambda \int_0^s (t-s) y(t) \, dt .$$

Example 2.   Reduce the boundary value problem

$$y''(s) + \lambda P(s) y = Q(s) , \tag{4}$$

$$y(a) = 0 , \qquad y(b) = 0 \tag{5}$$

to a Fredholm integral equation.

Comparing (4) and (5) with the notation of Section 5.2, we have $A = 0$, $B = \lambda P(s)$, $F(s) = Q(s)$, $y_0 = 0$, $y_1 = 0$. Substitution of these values in the relations (5.2.8) and (5.2.9) yields

$$f(s) = \int_a^s (s-t) Q(t) \, dt - [(s-a)/(b-a)] \int (b-t) Q(t) \, dt \tag{6}$$

and

$$K(s,t) = \begin{cases} \lambda P(t) [(s-a)(b-t)/(b-a)] , & s < t , \\ \lambda P(t) [(t-a)(b-s)/(b-a)] , & s > t , \end{cases} \tag{7}$$

which, when put in (5.2.7), gives the required integral equation. Note that the kernel is continuous at $s = t$.

As a special case of the above example, let us take the boundary value problem

$$y'' + \lambda y = 0 , \tag{8}$$

$$y(0) = 0 , \qquad y(\ell) = 0 . \tag{9}$$

Then, the relations (6) and (7) take the simple forms: $f(s) = 0$, and

$$K(s,t) = \begin{cases} (\lambda s/\ell)(\ell - t) , & s < t , \\ (\lambda t/\ell)(\ell - s) , & s > t . \end{cases} \tag{10}$$

Note that, although the kernels (7) and (10) are continuous at $s = t$, their derivatives are not continuous. For example, the derivative of the kernel (10) is

$$\partial K(s,t)/\partial s = \begin{cases} \lambda[1 - (t/\ell)] , & s < t , \\ -\lambda t/\ell , & s > t . \end{cases}$$

The value of the jump of this derivative at $s = t$ is

$$\left[\frac{dK(s,t)}{ds} \right]_{t+0} - \left[\frac{dK(s,t)}{ds} \right]_{t-0} = -\lambda .$$

Similarly, the value of the jump of the derivative of the kernel (7) at $s = t$ is

$$\left[\frac{dK(s,t)}{ds} \right]_{t+0} - \left[\frac{dK(s,t)}{ds} \right]_{t-0} = -\lambda P(t) .$$

Example 3. *Transverse oscillations of a homogeneous elastic bar.* Consider a homogeneous elastic bar with linear mass density d. Its axis coincides with the segment $(0, \ell)$ of the s axis when the bar is in its state of rest. It is clamped at the end $s = 0$, free at the end $s = \ell$, and is forced to perform simple harmonic oscillations with period $2\pi/\omega$. The problem, illustrated in Figure 5.1, is to find the deflection $y(s)$ that is parallel to the y axis and satisfies the system of equations

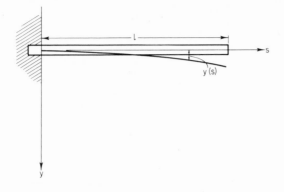

Figure 5.1

$$\frac{d^4y}{ds^4} - k^4 y = 0, \qquad k^4 = \frac{\omega^2 d}{EI}, \tag{11}$$

$$y(0) = y'(0) = 0, \tag{12}$$

$$y''(\ell) = y'''(\ell) = 0, \tag{13}$$

where EI is the bending rigidity of the bar.

The differential equation (11) with the initial conditions (12) can be reduced to the solution of a Volterra integral equation if we stipulate that

$$y''(0) = C_2, \qquad y'''(0) = C_3, \tag{14}$$

and subsequently determine the constants C_2 and C_3 with the help of (13). Indeed, when we compare the initial value problem embodied in (11), (12), and (14) with the system (5.1.9)–(5.1.15), we obtain the required integral equation

$$g(s) = k^4 \left(\frac{s^2}{2!} C_2 + \frac{s^3}{3!} C_3\right) + k^4 \int_0^s \frac{(s-t)^3}{3!} g(t) \, dt, \tag{15}$$

where

$$g(s) = d^4 y / ds^4. \tag{16}$$

The solution $y(s)$ of the differential equation (11) is

$$y(s) = \int_0^s \frac{(s-t)^3}{3!} \, g(t) \, dt + \frac{s^3}{3!} C_3 + \frac{s^2}{2!} C_2 \,. \tag{17}$$

We leave it to the reader to apply the conditions (13), determine the constants C_2 and C_3, and thereby complete the transformation of the system (11)–(13) into an integral equation.

The kernel of the integral equation (15) is that of convolution type and this equation can be readily solved by Laplace transform methods as explained in Chapter 9 (see Exercise 11 at the end of that chapter).

5.4. DIRAC DELTA FUNCTION

In physical problems, one often encounters idealized concepts such as a unit force acting at a point $s = s_0$ only, or an impulsive force acting at time $t = t_0$ only. These forces are described by the Dirac delta function $\delta(s-s_0)$ or $\delta(t-t_0)$ such that

$$\delta(x-x_0) = \begin{cases} 0 \,, & x \neq x_0 \,, \\ \infty \,, & x = x_0 \,, \end{cases} \tag{1}$$

where x stands for s in the case of a unit force and for t in the case of an impulsive force. Also,

$$\int \delta(x-x_0) \, dx = \begin{cases} 0 & \text{if } x_0 \text{ is not in } (a,b) \,, \\ 1 & \text{if } x_0 \text{ is in } (a,b) \,. \end{cases} \tag{2}$$

This function is supposed to have the sifting property

$$\int_{-\infty}^{\infty} \delta(x-x_0) \, \phi(x) \, dx = \phi(x_0) \tag{3}$$

for every continuous function $\phi(x)$.

The Dirac delta function has been successfully used in the description of concentrated forces in solid and fluid mechanics, point masses in the theory of gravitational potential, point charges in electrostatics, impulsive forces in acoustics, and various similar phenomena in other branches of physics and mechanics. In spite of the fact that scientists have used this function with success, the language of classical mathematics is inadequate to justify such a function. It is usually visualized as a limit of piecewise continuous functions $f(x)$ such as

$$f(x) = \begin{cases} 0, & 0 \leqslant x < x_0 - \frac{1}{2}\varepsilon, \\ p, & |x - x_0| \leqslant \frac{1}{2}\varepsilon, \\ 0, & x_0 + \frac{1}{2}\varepsilon < x < \ell, \end{cases} \qquad (4)$$

or as a limit of a sequence of suitable functions such as

$$f_k(x) = \begin{cases} k, & 0 < |x| < 1/k, \\ 0, & \text{for all other } x, \end{cases} \qquad (5)$$

where $k = 1, 2, 3, \ldots$, and

$$f_k(x) = \frac{1}{\pi} \frac{\sin kx}{x}. \qquad (6)$$

Our aim in this and the next chapter is to determine integral representation formulas for the solutions of linear ordinary and partial differential equations in such a manner so as to include the boundary conditions explicitly. To accomplish this task, we have to solve differential equations whose inhomogeneous term is a concentrated source. This is best done by introducing the theory of distributions, but that is not on our agenda. We shall, therefore, content ourselves with the above-mentioned properties of the delta function. Furthermore, we shall need the Heaviside function $H(x)$:

$$H(x) = \begin{cases} 0, & x < 0 \\ 1, & x > 0 \end{cases} \qquad (7)$$

and the relation

$$dH(x)/dx = \delta(x). \qquad (8)$$

5.5. GREEN'S FUNCTION APPROACH

We shall consider the initial and boundary value problems of Sections 5.1 and 5.2 in a different context. Let L be the differential operator

$$Lu(s) = [A(s)\frac{d^2}{ds^2} + B(s)\frac{d}{ds} + C(s)]u(s), \quad a < s < b, \tag{1}$$

where $A(s)$ is continuously differentiable, positive function. Its adjoint operator M is defined as

$$Mv(s) = \frac{d^2}{ds^2}[A(s)v(s)]$$

$$- \frac{d}{ds}[B(s)v(s)] + C(s)v(s), \quad a < s < b. \tag{2}$$

It follows by integration by parts that

$$\int (vLu - uMv)\, ds = [A(vu' - uv') + uv(B - A')]_a^b. \tag{3}$$

This is known as Green's formula for the operator L.

It is traditionally proved in the theory of ordinary differential equations that the relation

$$A(s)y'' + B(s)y' + C(s)y = \Phi(s) \tag{4}$$

can be converted to the form

$$\frac{d}{ds}\left[p(s)\frac{dy}{ds}\right] + q(s)y = F(s), \tag{5}$$

which is clearly self-adjoint. The function $p(s)$ is again continuously differentiable and positive and $q(s)$ and $F(s)$ are continuous in a given interval (a, b). Green's formula (3) for this operator takes the simple form

$$\int (vLu - uLv)\, ds = [p(s)(vu' - uv')]_a^b. \tag{6}$$

The homogeneous second-order equation

$$\frac{d}{ds}\left(p\frac{dy}{ds}\right) + qy = 0 \tag{7}$$

has exactly two linearly independent solutions $u(s)$ and $v(s)$ which are twice continuously differentiable in the interval $a < s < b$. Any other solution of this equation is a linear combination of $u(s)$ and $v(s)$, i.e., $y(s) = c_1 u(s) + c_2 v(s)$, where c_1 and c_2 are constants.

Initial Value Problems

Let us first consider the initial value problem

$$\frac{d}{ds}\left(p \frac{dy}{ds} \right) + qy = F(s) , \tag{8}$$

$$y(a) = 0 , \qquad y'(a) = 0 . \tag{9}$$

To formulate this problem into an integral equation, we consider the function

$$w(s) = u(s) \int_a^s v(t) F(t) \, dt - v(s) \int_a^s u(t) F(t) \, dt , \tag{10}$$

where u and v are solutions of the homogeneous equation (7) as mentioned above. The relation (10), when differentiated, gives

$$w'(s) = u'(s) \int_a^s v(t) F(t) \, dt - v'(s) \int_a^s u(t) F(t) \, dt$$

$$+ u(s) v(s) F(s) - u(s) v(s) F(s)$$

$$= u'(s) \int_a^s v(t) F(t) \, dt - v'(s) \int_a^s u(t) F(t) \, dt .$$

Hence, $w(a) = w'(a) = 0$, and

$$\frac{d}{ds}\left[p(s) \frac{dw}{ds} \right] = \frac{d}{ds}\left[p(s) \frac{du}{ds} \right] \int_a^s v(t) F(t) \, dt$$

$$- \frac{d}{ds}\left[p(s) \frac{dv}{ds} \right] \qquad \text{(equation continued)}$$

$$\times \int_a^s u(t)\,F(t)\,dt + p(s)\,[u'(s)\,v(s) - v'(s)\,u(s)]\,F(s)$$

$$= -q(s)\,w(s) + p(s)\,[u'(s)\,v(s) - v'(s)\,u(s)]\,F(s)\,, \qquad (11)$$

where we have used the fact that $u(s)$ and $v(s)$ satisfy the equation (7). In addition [dropping the argument (s) for p, u, v],

$$\frac{d}{ds}\,\{p(u'\,v - v'\,u)\} = \frac{d}{ds}\,(pu')\,v - \frac{d}{ds}\,(pv')\,u + pu'\,v' - pv'\,u' = 0\,,$$

also because u and v satisfy (7). This means that

$$p(s)\,[u'(s)\,v(s) - v'(s)\,u(s)] = A\,, \qquad (12)$$

where A is a constant. The negative of the expression in the brackets in the above relation is called the Wronskian $W(u, v; s)$ of the solutions u and v:

$$W(u, v; s) = u(s)\,v'(s) - v(s)\,u'(s)\,. \qquad (13)$$

From the relations (11) and (12), it follows that the function w as given by (10) satisfies the system

$$\frac{d}{ds}\left(p\frac{dw}{ds}\right) + qw = AF(s)\,, \qquad (14)$$

$$w(a) = 0\,, \qquad w'(a) = 0\,. \qquad (15)$$

Dividing (14) by the constant A and comparing it with (5), we derive the required relation $y(s)$ as

$$y(s) = \int_a^s R(s, t)\,F(t)\,dt\,, \qquad (16)$$

where

$$R(s, t) = (1/A)\,\{u(s)\,v(t) - v(s)\,u(t)\}\,. \qquad (17)$$

Note that $R(s, t) = -R(t, s)$.

It is easily verified that, for a fixed value of t, the function $R(s, t)$ is completely characterized as the solution of the initial value problem

$$LR = \frac{d}{ds}\left[p(s)\frac{dR}{ds}\right] + q(s)\,R = \delta(s-t)\,,$$

$$R\big|_{s=t} = 0 , \qquad \frac{dR}{ds}\bigg|_{s=t} = \frac{1}{p(t)} . \tag{18}$$

This function describes the influence on the value of y at s due to a concentrated disturbance at t. It is called the influence function. The function $G(s;t)$,

$$G(s;t) = \begin{cases} 0 , & s < t , \\ R(s,t) , & s > t \end{cases} \tag{19}$$

is called the causal Green's function.

Example.   Consider the initial value problem

$$y'' + y = F(s) , \qquad 0 < s < 1 , \qquad y(0) = y'(0) = 0 . \tag{20}$$

The influence function $R(s,t)$ is the solution of the system

$$\frac{d^2 R}{ds^2} + R = \delta(s-t) , \qquad R\big|_{s=t} = 0 , \qquad \frac{dR}{ds}\bigg|_{s=t} = 1 . \tag{21}$$

The required value of R, clearly, is $R(s,t) = \sin(s-t)$, and the integral representation formula for the initial value problem (20) is

$$y(s) = \int_a^s \sin(s-t) F(t) \, dt . \tag{22}$$

When the values of $y(a)$ and $y'(a)$ are prescribed to be other than zero, then we simply add a suitable solution $c_1 u + c_2 v$ of (7) to the integral equation (16) and evaluate the constants c_1 and c_2 by the prescribed conditions. For example,

$$y'' + y = F(s) , \qquad y(0) = 1 , \qquad y'(0) = -1 \tag{23}$$

has the solution

$$y(s) = \int_0^s [\sin(s-t)] F(t) \, dt + c_1(\sin s) + c_2(\cos s) . \tag{24}$$

With the help of the prescribed conditions, we find that $c_1 = -1$ and $c_2 = 1$.

Boundary Value Problems

Let us now consider the boundary value problems and start with the simplest one,

$$\frac{d}{ds}\left(p \frac{dy}{ds} \right) + qy = F(s) , \qquad a \leqslant s \leqslant b , \tag{25}$$

$$y(a) = 0 , \qquad y(b) = 0 . \tag{26}$$

We attempt to write its general solution as an integral equation of the form

$$y(s) = \int_a^s R(s, t) F(t) \, dt + c_1 u(s) + c_2 v(s) , \tag{27}$$

where $u(s)$ and $v(s)$ are the solutions of the homogeneous equation (7). When we substitute the conditions (26) in (27), we obtain

$$c_1 u(a) + c_2 v(a) = 0 ,$$

$$c_1 u(b) + c_2 v(b) = - \int R(b, t) F(t) \, dt ,$$

which will determine a unique pair of constants c_1 and c_2 provided the following holds for the determinant D:

$$D = u(a) v(b) - v(a) u(b) \neq 0 ; \tag{28}$$

for the time being, we assume this to be true. Therefore,

$$c_1 = [v(a)/D] \int R(b, t) F(t) \, dt$$

$$= [v(a)/D] \int_a^s R(b, t) F(t) \, dt + [v(a)/D] \int_s^b R(b, t) F(t) \, dt , \tag{29}$$

$$c_2 = -[u(a)/D] \int R(b, t) F(t) \, dt$$

$$= -[u(a)/D] \int_a^s R(b, t) F(t) \, dt - [u(a)/D] \int_s^b R(b, t) F(t) \, dt . \tag{30}$$

Putting these values of c_1 and c_2 in the relation (27), we have the solution $y(s)$ as

$$y(s) = \int\limits_{a}^{s} \{R(s,t) + (1/D)[v(a)u(s) - u(a)v(s)]R(b,t)\} F(t)\, dt$$

$$+ \int\limits_{s}^{b} (1/D)[v(a)u(s) - u(a)v(s)]R(b,t)F(t)\, dt \,. \tag{31}$$

Using (12) and (17) and doing some algebraic manipulation, we find that

$$R(s,t) + \{[v(a)u(s) - u(a)v(s)]/D\} R(b,t)$$

$$= (1/AD)[u(a)v(t) - v(a)u(t)][u(s)v(b) - v(s)u(b)] \,. \tag{32}$$

Finally, we define the function $G(s;t)$:

$$G(s;t) = \begin{cases} (1/AD)[u(s)v(a) - v(s)u(a)][u(t)v(b) - v(t)u(b)] \,, \\ \qquad s < t \,, \\ (1/AD)[u(t)v(a) - v(t)u(a)][u(s)v(b) - v(s)u(b)] \,, \\ \qquad s > t \,. \end{cases} \tag{33}$$

Then the solution $y(s)$ as given by (29) takes the elegant form

$$y(s) = -\int G(s;t)F(t)\, dt \,. \tag{34}$$

The function $G(s;t)$ is called the Green's function. It is clearly symmetric:

$$G(s;t) = G(t;s) \,. \tag{35}$$

Furthermore, it satisfies, for all t, the following auxiliary problem:

$$LG = \frac{d}{ds}\left[p(s)\frac{dG}{ds}\right] + q(s)\,G = -\delta(s-t) \,, \tag{36}$$

$$G\big|_{s=a} = G\big|_{s=b} = 0 \,, \tag{37}$$

$$G\big|_{s=t+0} - G\big|_{s=t-0} = 0 \,, \tag{38}$$

$$\frac{dG}{ds}\bigg|_{t+0} - \frac{dG}{ds}\bigg|_{t-0} = -\frac{1}{p(t)} \,, \tag{39}$$

where by $G\big|_{s=t+0}$ we mean the limit of $G(s;t)$ as s approaches t from the right, and there are similar meanings for the other expressions. Thus, the condition (38) implies that the Green's function is continuous at $s = t$. Similarly, the condition (39) states that dG/ds has a jump discontinuity of magnitude $-\{1/p(t)\}$ at $s = t$. The conditions (38) and (39) are called the matching conditions.

It is instructive to note that the relation (39) is a consequence of the relations (35) and (36). Indeed, the value of the jump in the derivative of $G(s;t)$ can be obtained by integrating (36) over small interval $(t-\varepsilon, s)$ and by recalling that the indefinite integral of $\delta(s-t)$ is the Heaviside function $H(s-t)$. The result is

$$p(t)\frac{dG(s;t)}{ds} + \int_{t-\varepsilon}^{s} q(x)\, G(x;t)\, dx = p(t-\varepsilon)\frac{dG(t-\varepsilon;s)}{ds} - H(s-t) \ .$$

When s traverses the source point t, then on the right side the Heaviside function has a unit jump discontinuity. Since other terms are continuous functions of s, it follows that dG/ds has, at t, a jump discontinuity as given by (39).

Example.   Consider the boundary value problem

$$y'' = F(s) \ , \qquad y(0) = y(\ell) = 0 \ . \tag{40}$$

Comparing this system with the relations (25), (26), and (36)–(39), we readily evaluate the Green's function:

$$G(s;t) = \begin{cases} (s/\ell)(\ell - t) \ , & s < t \ , \\ (t/\ell)(\ell - s) \ , & s > t \ , \end{cases} \tag{41}$$

which is the kernel (5.3.10) except for the factor λ. The solution of (40) now follows by substituting this value in the equation (34). Incidentally, by introducing the notation

$$s_< = \begin{cases} s \ , & s \leqslant t \ , \\ t \ , & s \geqslant t \ , \end{cases} \qquad \text{or} \quad \min(s,t) \ ,$$

and

$$s_> = \begin{cases} t \ , & s \leqslant t \ , \\ s \ , & s \geqslant t \ , \end{cases} \qquad \text{or} \quad \max(s,t) \ ,$$

the relation (41) takes the compact form

$$G(s;t) = (1/\ell)[s_<(\ell - s_>)], \qquad 0 \leqslant s, t \leqslant \ell. \tag{42}$$

It follows from the properties (36)–(39) of the Green's function $G(s;t)$ that $\partial G(s;a)/\partial t$ satisfies the system of equations

$$\frac{d}{ds}\left[p(s)\frac{d}{ds}\left\{\frac{\partial G(s;a)}{\partial t}\right\}\right] + q(s)\frac{\partial G(s;a)}{\partial t} = 0, \qquad a < s < b,$$

$$\frac{\partial G(a;a)}{\partial t} = \frac{1}{p(a)}, \qquad \frac{\partial G(b;a)}{\partial t} = 0. \tag{43}$$

Similarly, $\partial G(s;b)/\partial t$ satisfies the system

$$\frac{d}{ds}\left[p\frac{d}{ds}\left\{\frac{\partial G(s;b)}{\partial t}\right\}\right] + q(s)\frac{\partial G(s;b)}{\partial t} = 0, \qquad a < s < b,$$

$$\frac{\partial G(a;b)}{\partial t} = 0, \qquad \frac{\partial G(b;b)}{\partial t} = -\frac{1}{p(b)}. \tag{44}$$

Hence, the boundary value problem

$$(py')' + qy = F, \qquad y(a) = \alpha, \qquad y(b) = \beta \tag{45}$$

has the solution

$$y(s) = -\int G(s;t)F(t)\,dt$$

$$+ \alpha p(a)\frac{\partial G(s;a)}{\partial t} - \beta p(b)\frac{\partial G(s;b)}{\partial t}, \tag{46}$$

as is easily verified.

Finally, we present the integral-equation formulation for the boundary value problem with more general and inhomogeneous end conditions:

$$(py')' + qy = F(s),$$

$$-\mu_1 y'(a) + \nu_1 y(a) = \alpha, \qquad \mu_2 y'(b) + \nu_2 y(b) = \beta. \tag{47}$$

When we proceed to solve this system in the same way as we did the system (1)–(2), the Green's function for the present case can also be derived provided the determinant

$$D = [-\mu_1 u'(a) + \nu_1 u(a)][\mu_2 v'(b) + \nu_2 v(b)]$$

$$- [-\mu_1 v'(a) + \nu_1 v(a)][\mu_2 u'(b) + \nu_2 u(b)] \neq 0.$$

Indeed, $G(s;t)$ possesses the following properties:

$$LG = \frac{d}{ds}\left[p(s)\frac{dG}{ds}\right] + q(s)\,G = -\delta(s-t)$$

$$-\mu_1 \frac{dG}{ds}\bigg|_{s=a} + v_1\,G\big|_{s=a} = \mu_2 \frac{dG}{ds}\bigg|_{s=b} + v_2\,G\big|_{s=b} = 0\ .$$

$$G\big|_{s=t+0} - G\big|_{s=t-0} = 0\ ,$$

$$\frac{dG}{ds}\bigg|_{s=t+0} - \frac{dG}{ds}\bigg|_{s=t-0} = -\frac{1}{p(t)}\ , \tag{48}$$

and the condition of symmetry. With the help of the Green's function, the boundary value problem (47) has the unique solution

$$y(s) = -\int G(s;t)\,F(t)\,dt + \frac{p(a)}{\mu_1}\,\alpha G(s;a) + \frac{p(b)}{\mu_2}\,\beta G(s;b)\ , \tag{49}$$

provided μ_1 and μ_2 do not vanish. If $\mu_1 = 0$, then the factor $(1/\mu_1)\,G(s;a)$ is replaced by $(1/v_1)\,\partial G(s;a)/\partial t$. By the same token, if $\mu_2 = 0$, we replace $(1/\mu_2)\,G(s;b)$ by $-(1/v_2)\,\partial G(s;b)/\partial t$. In view of the relation $D \neq 0$, we cannot have both μ_1 and v_1 or both μ_2 and v_2 equal to zero. When α and β are zero, the relation (49) reduces to (34).

The Sturm–Liouville problem consists in solving a differential equation of the form

$$(py')' + qy + \lambda ry = F(s) \tag{50}$$

involving a parameter λ and subject to a pair of homogeneous boundary conditions

$$-\mu_1 y'(a) + v_1 y(a) = 0\ , \qquad \mu_2 y'(b) + v_2 y(b) = 0\ . \tag{51}$$

The values of λ for which this problem has a nontrivial solution are called the eigenvalues. The corresponding solutions are the eigenfunctions. In case $p(a) = p(b)$, the boundary conditions (51) are replaced by the periodic boundary conditions

$$y(a) = y(b)\ , \qquad y'(a) = y'(b)\ .$$

From formula (49), it follows that the solution of equation (50) subject to the conditions (51) is

$$y(s) = \lambda \int r(t) G(s;t) y(t) \, dt - \int G(s;t) F(t) \, dt , \qquad (52)$$

which is a Fredholm integral equation of the second kind. In this equation, $G(s;t) r(t)$ is not symmetric unless the function r is a constant. However, by setting

$$[r(s)]^{1/2} y(s) = Y(s) ,$$

under the assumption that $r(s)$ is nonnegative over the interval (a,b), the equation (52) takes the form

$$y(s)[r(s)]^{1/2} = \lambda \int G(s;t) [r(s)]^{1/2} [r(t)]^{1/2} [r(t)]^{1/2} y(t) \, dt$$

$$- \int G(s;t) [r(s)]^{1/2} [r(t)]^{1/2} \frac{F(t)}{[r(t)]^{1/2}} \, dt ,$$

or

$$Y(s) = \lambda \int \tilde{G}(s;t) Y(t) \, dt - \int \tilde{G}(s;t) \frac{F(t)}{[r(t)]^{1/2}} \, dt , \qquad (53)$$

where $\tilde{G}(s;t) = G(s;t) [r(s)]^{1/2} [r(t)]^{1/2}$ is a symmetric kernel.

The above discussion on boundary value problems is based on the assumption that $D = u(a)v(b) - v(a)u(b)$ does not vanish. If it vanishes, then the homogeneous equations

$$c_1 u(a) + c_2 v(a) = 0 , \qquad c_1 u(b) + c_2 v(b) = 0$$

have a nontrivial solution (c_1, c_2), and the function $w(s) = c_1 u(s) + c_2 v(s)$ satisfies the completely homogeneous system

$$(pw')' + qw = 0 , \qquad w(a) = w(b) = 0 . \qquad (54)$$

Therefore, if y is a solution of (25), (45), or (47), then so is $y + cw$ for any constant c. This means these systems do not have a unique solution. This is not all. There is an additional consistency condition which must be satisfied for these systems to have a solution. Take, for example, the system (45). Multiply the differential equation (45) by w and integrate from a to b and get

$$\int w(s) F(s) \, ds = \int w(s) [(py')' + qy] \, ds$$

$$= [wpy' - w'py]_a^b + \int y[(pw')' + qw] \, ds$$

$$= p(a) w'(a) \alpha - p(b) w'(b) \beta . \qquad (55)$$

Therefore, if (45) is to have a solution, then the given function $F(x)$ must satisfy the consistency condition (55). For $\alpha = \beta = 0$, we get the consistency condition

$$\int w(s) F(s)\, ds = 0 , \tag{56}$$

for the system (25). Thus, if D is zero, then we either have no solution or many solutions; but never just one.

5.6. EXAMPLES

Example 1.   Reduce the boundary value problem

$$y'' + \lambda y = 0 , \tag{1}$$

$$y(0) = 0 , \qquad y'(1) + v_2 y(1) = 0 , \tag{2}$$

to a Fredholm integral equation.

From the properties $(48)_1$ and $(48)_2$, we must have

$$G(s;t) = \begin{cases} A_1(t)s , & s < t , \\ A_2(t)[1 + v_2(1-s)] , & s > t . \end{cases}$$

The consequence of the symmetry of the Green's function is

$$A_1 = C[1 + v_2(1-t)] , \qquad A_2 = Ct ,$$

where C is a constant independent of t. The jump condition $(48)_4$ yields

$$Ct(-v_2) - C[1 + v_2(1-t)] = -1$$

or

$$C = 1/(1+v_2) .$$

Thus, the Green's function is completely determined:

$$G(s;t) = \begin{cases} [1 + v_2(1-t)]s/(1+v_2) , & s < t , \\ [1 + v_2(1-s)]t/(1+v_2) , & s > t . \end{cases} \tag{3}$$

The required integral equation is

$$y(s) = \lambda \frac{1 + v_2(1-s)}{1 + v_2} \int_0^s t y(t) \, dt$$

$$+ \lambda \frac{s}{1 + v_2} \int_s^1 [1 + v_2(1-t)] y(t) \, dt + \frac{s}{1 + v_2} \, . \tag{4}$$

Example 2. Reduce the Bessel equation

$$s^2 \frac{d^2 y}{ds^2} + s \frac{dy}{ds} + (\lambda s^2 - 1) y = 0 \tag{5}$$

with end conditions

$$y(0) = 0, \qquad y(1) = 0, \tag{6}$$

to a Fredholm integral equation.

The differential equation (5) can be written as

$$(sy')' + [(-1/s) + \lambda s] y = 0 \, . \tag{7}$$

Comparing (7) with (5.5.50), we obtain

$$p(s) = s, \qquad q(s) = -1/s, \qquad r(s) = s, \qquad F(s) = 0 \, .$$

To find the Green's function, we observe that the two linearly independent solutions of the equation

$$s^2 \frac{d^2 y}{ds^2} + s \frac{dy}{ds} - y = 0$$

are s and $1/s$. Therefore, from (5.5.36)–(5.5.39), it follows that

$$G(s;t) = \begin{cases} (s/2t)(1-t^2), & s < t, \\ (t/2s)(1-s^2), & s > t. \end{cases} \tag{8}$$

Substituting this value of G in the relation (5.5.52), we get the required Fredholm integral equation.

Example 3. Reduce the following boundary value problem to a Fredholm integral equation:

$$y'' + \lambda s y = 1, \qquad y(0) = 0, \qquad y(\ell) = 1 \, . \tag{9}$$

The Green's function is the same as given by (5.5.41):

$$G(s;t) = \begin{cases} (s/\ell)(\ell - t), & s < t, \\ (t/\ell)(\ell - s), & s > t. \end{cases} \tag{10}$$

The expression for the integral equation then follows from (5.5.46):

$$y(s) = -\int_0^\ell G(s;t)\,dt + \lambda \int_0^\ell G(s;t)\,ty(t)\,dt - \partial G(s;\ell)/\partial t. \tag{11}$$

The function $\partial G(s;\ell)/\partial t$ satisfies the system of equations (5.5.44), which for the present case become

$$\frac{d^2}{ds^2}\left\{\frac{\partial G(s;\ell)}{\partial t}\right\} = 0, \qquad \frac{\partial G(0;\ell)}{\partial t} = 0, \qquad \frac{\partial G(\ell;\ell)}{\partial t} = -1, \tag{12}$$

whose solution is

$$\partial G(s;\ell)/\partial t = -s/\ell. \tag{13}$$

Substituting (12) and (13) in (11), we have

$$y(s) = (s/\ell) - \int_0^s (t/\ell)(\ell - s)\,dt - \int_s^\ell (s/\ell)(\ell - t)\,dt$$

$$+ \lambda \int_0^\ell G(s;t)\,ty(t)\,dt$$

or

$$y(s) = (s/2\ell)[2 - \ell^2 + s\ell] + \lambda \int_0^\ell G(s;t)\,ty(t)\,dt. \tag{14}$$

5.7. GREEN'S FUNCTION FOR NTH-ORDER ORDINARY DIFFERENTIAL EQUATION

The boundary value problem consisting of a differential equation of order n and prescribed end conditions can also be transformed to a Fredholm integral equation in the manner of Section 5.5. For instance, take the boundary value problem

$$\frac{d^2}{ds^2}\left\{x(s)\frac{d^2 y}{ds^2}\right\} + \frac{d}{ds}\left\{p(s)\frac{dy}{ds}\right\} + q(s)y - \lambda r(s)y = F(s),\qquad (1)$$

where λ is a parameter and the boundary conditions are

$$(xy'')' + py' = 0,\qquad \text{or}\quad y\quad \text{prescribed},$$
$$xy'' = 0,\qquad \text{or}\quad y'\quad \text{prescribed}.\qquad (2)$$

The Green's function $G(s;t)$ for the problem (1)–(2) is such that it satisfies the differential equation

$$\frac{d^2}{ds^2}\left\{x(s)\frac{d^2 y}{ds^2}\right\} + \frac{d}{ds}\left\{p(s)\frac{dy}{ds}\right\} = \delta(s-t),\qquad (3)$$

together with the prescribed homogeneous boundary conditions. In addition, G, $\partial G/\partial s$, and $\partial^2 G/\partial s^2$ are continuous at $s = t$. The value of the jump in the third derivative of G is

$$\left.\frac{\partial^3 G}{\partial s^3}\right|_{s=t+0} - \left.\frac{\partial^3 G}{\partial s^3}\right|_{s=t-0} = \frac{1}{x(s)}.\qquad (4)$$

Finally, G is symmetric.

In terms of the Green's function with the above five properties, the boundary value problem (1)–(2) reduces to the integral equation

$$y(s) = \lambda \int G(s,t)y(t)\, dt - \int G(s,t)F(t)\, dt.\qquad (5)$$

Example.   Consider the boundary value problem

$$\frac{d^4 y}{ds^4} + \lambda y = -f(s),\qquad (6)$$

$$y(0) = 0 = y'(0),\qquad y(1) = 0 = y'(1).\qquad (7)$$

The homogeneous equation

$$d^4 y/ds^4 = 0,$$

has the four linearly independent solutions $1, s, s^2, s^3$. Therefore, we take the value of $G(s;t)$ to be

$$G(s;t) = \begin{cases} A_0(t) + A_1(t)s + A_2(t)s^2 + A_3(t)s^3, & s < t, \\ B_0(t) + B_1(t)s + B_2(t)s^2 + B_3(t)s^3, & s > t. \end{cases}\qquad (8)$$

The boundary conditions at the end points give

$$A_0(t) = 0, \qquad A_1(t) = 0, \qquad B_2 = -3B_0 - 2B_1, \qquad B_3 = 2B_0 + B_1.$$

Thus, the relation (8) becomes

$$G(s;t) = \begin{cases} A_2(t)s^2 + A_3 s^3, & s < t, \\ (1-s)^2[B_0(t)(1+2s) + B_1 s], & s > t. \end{cases}$$

The remaining constants are determined by applying the matching conditions at $s = t$, which result in the simultaneous equations

$$t^2 A_2 + t^3 A_3 - (1 - 3t^2 + 2t^3) B_0 - t(1-t)^2 B_1 = 0,$$

$$2t A_2 + 3t^2 A_3 + 6t(1-t) B_0 - (1 - 4t + 3t^2) B_1 = 0,$$

$$2A_2 + 6t A_3 + 6(1-2t) B_0 + 2(2-3t) B_1 = 0,$$

$$6A_3 - 12B_0 - 6B_1 = 1,$$

whose solution is

$$A_2(t) = -\tfrac{1}{2}t(1-t)^2, \qquad A_3(t) = \tfrac{1}{6}(1-t)^2(2t+1),$$

$$B_0(t) = \tfrac{1}{6}t^3, \qquad B_1(t) = -\tfrac{1}{2}t^2.$$

Hence,

$$G(s;t) = \begin{cases} \tfrac{1}{6}s^2(1-t)^2(2st+s-3t), & s < t, \\ \tfrac{1}{6}t^2(1-s)^2(2st+t-3s), & s > t. \end{cases} \tag{9}$$

The required Fredholm integral equation then follows by substituting this value in the relation (5).

5.8. MODIFIED GREEN'S FUNCTION

We observed at the end of the Section 5.5 that, if the homogeneous equation $Ly = 0$, where L is the self-adjoint operator as defined in equation (5.5.5), with prescribed end conditions, has a nontrivial solution $w(s)$, then the corresponding inhomogeneous equation either has no solution or many solutions, depending on the consistency condition. This means that the Green's function, as defined in Section 5.5, does not exist, because

$$\int \delta(s-t)\,w(s)\,ds \neq 0\,, \qquad a < t < b\,. \tag{1}$$

A method of constructing the Green's function for this type of problem is now presented. Such a function is called the modified Green's function and we shall denote it by $G_M(s;t)$. We start by choosing a normalized solution of the completely homogeneous system so that $\int w^2(s)\,ds = 1$. The modified Green's function is to satisfy the following properties.

(a) G_M satisfies the differential equation

$$LG_M(s;t) = \delta(s-t) - w(s)w(t)\,. \tag{2}$$

This amounts to introducing an additional source density so that the consistency condition is satisfied. Indeed,

$$\int [\delta(s-t) - w(s)w(t)]\,w(s)\,ds = 0\,.$$

(b) G_M satisfies the prescribed homogeneous end conditions.

(c) G_M is continuous at $s = t$.

(d) G_M satisfies

$$\left.\frac{dG_M}{ds}\right|_{t+0} - \left.\frac{dG_M}{ds}\right|_{t-0} = -\frac{1}{p(t)}\,. \tag{3}$$

Thus, the construction is similar to that for the ordinary Green's function except that the modified Green's function is not uniquely determined; we can add $cw(s)$ to it without violating any of the above four properties. It is often convenient to choose a particular modified Green's function that is a symmetric function of s and t. This is accomplished by defining two functions $G_M(s;t_1)$ and $G_M(s;t_2)$ which satisfy the equations

$$LG_M(s;t_1) = \delta(s-t_1) - w(t_1)w(s)\,, \tag{4}$$

$$LG_M(s;t_2) = \delta(s-t_2) - w(t_2)w(s)\,, \tag{5}$$

along with prescribed (same for both) homogeneous end conditions.

Multiply (4) by $G_M(s; t_2)$ and (5) by $G_M(s; t_1)$, subtract, and integrate from a to b. Finally, use the Green's formula (5.5.6) and get

$$G_M(t_1; t_2) - G_M(t_2; t_1) + w(t_1) \int G_M(s; t_2) w(s) \, ds$$
$$- w(t_2) \int G_M(s; t_1) w(s) \, ds = 0 . \tag{6}$$

Now, if we impose the condition:
(e) G_M satisfies the property

$$\int G_M(s; t) w(s) \, ds = 0 , \tag{7}$$

then it follows from (6) that G_M will be symmetric. Thereby, the Green's function is uniquely defined.

Finally, we can reduce the inhomogeneous equation

$$Ly = F(s) \tag{8}$$

with prescribed homogeneous end conditions into an integral equation when the consistency condition $\int F(s) w(s) \, ds = 0$, is satisfied. Indeed, by following the usual procedure, one gets

$$\int (G_M Ly - yLG_M) \, ds = \int G_M(s; t) F(s) \, ds - y(t)$$
$$+ \int y(s) w(s) w(t) \, ds . \tag{9}$$

But the left side is zero because of the Green's formula (5.5.6) and the end conditions. Thus, we are left with the relation

$$y(t) = \int G_M(s; t) F(s) \, ds + Cw(t) , \tag{10}$$

where $C = \int y(s) w(s) \, ds$, is a constant, although as yet undetermined. By interchanging s and t in (10), we have

$$y(s) = \int G_M(t; s) F(t) \, dt + Cw(s) . \tag{11}$$

For a symmetric Green's function, the above result becomes

$$y(s) = \int G_M(s; t) F(t) \, dt + Cw(s) . \tag{12}$$

When $w(s) = 0$, this reduces to the relation (5.5.34) [note that G in (5.5.34) is negative of the G_M in (12)].

5.9. EXAMPLES

Example 1. Transform the boundary value problem

$$-\left(\frac{d^2 y}{ds^2} + \lambda y\right) = F(s), \qquad y'(0) = y'(\ell) = 0, \qquad 0 \leqslant s \leqslant \ell \quad (13)$$

into an integral equation.

The self-adjoint operator $-d^2 y/ds^2 = 0$, $0 \leqslant s \leqslant \ell$, with the end conditions $y(0) = y'(\ell) = 0$ has the nontrivial solution $y = \text{const}$. We take it to be $1/\ell^{1/2}$ because $\int_0^\ell (1/\ell)\, ds = 1$. We therefore have to solve the system

$$-\frac{d^2 G_M(s;t)}{ds^2} = \delta(s-t) - \frac{1}{\ell}, \qquad \frac{dG(0;t)}{ds} = \frac{dG(\ell;t)}{ds} = 0. \quad (14)$$

The solution of the system (14) for $s < t$ that satisfies the end condition at $s = 0$ is $A + (s^2/2\ell)$. Similarly, the solution that satisfies the end condition at $s = \ell$ is $B - s + (s^2/2\ell)$. Hence,

$$G_M(s;t) = \begin{cases} A + (s^2/2\ell), & s < t, \\ B - s + (s^2/2\ell), & s > t. \end{cases}$$

The condition of continuity at $s = t$ implies that $B = A + t$, while the jump condition on dG_M/ds is automatically satisfied. Thus,

$$G_M(s;t) = \begin{cases} A + (s^2/2\ell), & s < t, \\ A + t - s + (s^2/2\ell), & s > t. \end{cases}$$

The constant A is determined by the symmetry condition (7):

$$\int_0^t \left(A + \frac{s^2}{2\ell}\right) ds + \int_t^\ell \left(A + t - s + \frac{s^2}{2\ell}\right) ds = 0$$

or

$$A = \frac{1}{\ell}\left(\frac{\ell^2}{3} + \frac{t^2}{2} - t\ell\right).$$

Thereby, the symmetric Green's function is completely determined:

$$G_M(s;t) = \frac{\ell}{3} + \frac{1}{2\ell}(s^2+t^2) - \begin{cases} t, & s < t, \\ s, & s > t. \end{cases} \tag{15}$$

The expression for the integral equation follows from the formula (5.8.12),

$$y(s) = -\lambda \int_0^\ell G_M(s;t)y(t)\,dt + \int_0^\ell G_M(s;t)F(s)\,ds + (1/\ell)\int_0^\ell y(s)\,ds. \tag{16}$$

Example 2. Transform the boundary value problem

$$\frac{d}{ds}\left\{(1-s^2)\frac{dy}{ds}\right\} + \lambda y = 0, \qquad y(-1), \qquad y(1) \quad \text{finite} \tag{17}$$

into an integral equation.

The operator $-(d/ds)[(1-s^2)\,dy/ds]$ is a self-adjoint Legendre operator. The function $w(s) = 1/\sqrt{2}$ satisfies this operator as well as the boundary conditions. Hence, we have to solve the equation

$$-\frac{d}{ds}\left\{(1-s^2)\frac{dG_M}{ds}\right\} = \delta(s-t) - \tfrac{1}{2}. \tag{18}$$

For $s \neq t$, we have

$$(1-s^2)\frac{dG_M}{ds} = \frac{s}{2} + A$$

or

$$\frac{dG_M}{ds} = \frac{s}{2(1-s^2)} + \frac{A}{1-s^2}$$

or

$$G_M = -\tfrac{1}{4}\{\log[(1+s)(1-s)]\}$$
$$+ \tfrac{1}{2}A\{\log[(1+s)/(1-s)]\} + B. \tag{19}$$

Thus,

$$G_M(s;t) = \begin{cases} (\tfrac{1}{2}A - \tfrac{1}{4})\log(1+s) - (\tfrac{1}{2}A + \tfrac{1}{4})\log(1-s) + B, \\ \qquad s < t, \\ (\tfrac{1}{2}C - \tfrac{1}{4})\log(1+s) - (\tfrac{1}{2}c + \tfrac{1}{4})\log(1-s) + D, \\ \qquad s > t. \end{cases} \tag{20}$$

Since $G_M(s;t)$ has to be finite at -1 and $+1$, we must take $A = \frac{1}{2}$, $C = -\frac{1}{2}$, and the relation (20) reduces to

$$G_M(s;t) = \begin{cases} -\frac{1}{2}\log(1-s) + B, & s < t, \\ -\frac{1}{2}\log(1+s) + D, & s > t. \end{cases} \tag{21}$$

The continuity of G_M at $s = t$ implies that

$$B - D = -\frac{1}{2}\log(1+t) + \frac{1}{2}\log(1-t). \tag{22}$$

The jump condition at $s = t$ is automatically satisfied. Finally, the symmetry of G_M yields

$$0 = \int_{-1}^{1} G_M(s;t)\,ds = \int_{-1}^{t} [B - \tfrac{1}{2}\log(1-s)]\,ds + \int_{t}^{1} [D - \tfrac{1}{2}\log(1+s)]\,ds$$

or

$$B + D = 2(\log 2) - 1 - \tfrac{1}{2}[\log(1+t) + \log(1-t)]. \tag{23}$$

From (22) and (23), we obtain the values of B and D as

$$B = (\log 2) - \tfrac{1}{2} - \tfrac{1}{2}\log(1+t), \qquad D = (\log 2) - \tfrac{1}{2} - \tfrac{1}{2}\log(1-t). \tag{24}$$

Putting these values in (21), we have

$$G_M(s;t) = (\log 2) - \tfrac{1}{2} - \begin{cases} \frac{1}{2}\log[(1-s)(1+t)], & s < t, \\ \frac{1}{2}\log[(1+s)(1-t)], & s > t. \end{cases} \tag{25}$$

The required integral equation now follows by using the formula (5.8.12).

EXERCISES

Transform the following boundary value problems to integral equations:

1. $y'' + y = 0$, $y(0) = 0$, $y'(0) = 1$.

2. $y'' + xy = 1$, $y(0) = y(1) = 0$.

3. $y'' + y = s$, $y'(0) = y'(1) = 0$.

4. $y'' + y = s$, $y(0) = 1$, $y'(1) = 0$.

5. $y'' + (y'/2s) + \lambda s^{\frac{1}{2}} y = 0$, $y(0) = y(1) = 0$.

6. $y'' - y = f(s)$, $y(0) = y(1) = 0$.

7. Reduce the differential equation

$$y'' + \{P(s) - \eta^2\} y = 0,$$

η a known positive constant, with the end conditions $y(0) = 0$, $(y'/y) = -\eta$ at $s = s_0$, into a Fredholm integral equation.

8. Convert the initial value problem

$$y'' + \lambda s^2 y = 0,   y'(0) = 0 = y(1)$$

into a Volterra integral equation.

9. Find the boundary value problem that is equivalent to the integral equation

$$y(x) = \lambda \int_{-1}^{1} (1 - |x - t|) y(t) \, dt .$$

10. (a) Show that the Green's function for the Bessel operator of order n

$$Ly = (d/ds)(x \, dy/ds) - (n^2/s) y ,   n \neq 0$$

with the end conditions

$$y(0) = y(1) = 0$$

is

$$G(s; t) = \begin{cases} s^n/t^n(1 - t^{2n}) , & s < t , \\ t^n/s^n(1 - s^{2n}) , & s > t . \end{cases}$$

(b) Use the result of part (a) to reduce the problem

$$s^2 y'' + sy' + (\lambda s^2 - n^2) y = 0 ,   y(0) = y(1) = 0$$

to an integral equation.

11. Reduce the boundary value problem

$$\frac{d^2}{ds^2} \left\{ (4+s)^3 \frac{d^2 y}{ds^2} \right\} - \lambda(4+s) y = 0 ,$$

$$y(0) = 0 = y(1) , \qquad y''(0) = 0 = y'''(1)$$

to a Fredholm integral equation.

12. Extend the theory of Section 5.8 to the case when the completely homogeneous system has two linearly independent solutions $w_1(s)$ and $w_2(s)$.

13. Find the modified Green's function for the systems

$$y'' - \lambda y = 0 , \qquad y(0) = y(1) , \qquad y'(0) = y'(1)$$

14. Transform the boundary value problem

$$y'' + y = f(s) , \qquad y(0) = y(\pi) = 0 ,$$

into an integral equation.

Hint: The self-adjoint operator $-(y'' + y) = 0$ has the nontrivial solution $(2/\pi)^{1/2} \sin s$ and it satisfies the boundary conditions. Now proceed as in the examples in Section 5.9.

CHAPTER 6

APPLICATIONS TO PARTIAL DIFFERENTIAL EQUATIONS

6.1. INTRODUCTION

The applications of integral equations are not restricted to ordinary differential equations. In fact, the most important applications of integral equations arise in finding the solutions of boundary value problems in the theory of partial differential equations of the second order. The boundary value problems for equations of elliptic type can be reduced to Fredholm integral equations, while the study of parabolic and hyperbolic differential equations leads to Volterra integral equations. We shall confine our attention to the linear partial differential equations of the elliptic type, specifically, to the Laplace, Poisson, and Helmholtz equations, wherein lie the most interesting and important achievements of the theory of integral equations.

Three types of boundary conditions arise in the study of elliptic partial differential equations. The first type is the Dirichlet condition. In this case, we prescribe the value of the solution on the boundary. The second type is the Neumann condition. In this case, we prescribe the normal derivative of the solution on the boundary. When we prescribe the Dirichlet conditions on some parts and the Neumann conditions on other parts of the boundary, we have a mixed boundary value problem.

The divergence theorem and the two Green's identities will be used repeatedly in this chapter. They are as follows:

Divergence theorem:

$$\int_R \operatorname{div} \mathbf{A} \, dV = \int_S \mathbf{A} \cdot \mathbf{n} \, dS \; ; \tag{1}$$

Green's first identity:

$$\int_R u \nabla^2 v \, dV = -\int_R (\operatorname{grad} u \cdot \operatorname{grad} v) \, dV + \int_S u \frac{\partial v}{\partial n} \, dS \; ; \tag{2}$$

Green's second identity:

$$\int_R (u \nabla^2 v - v \nabla^2 u) \, dV = \int_S \left(u \frac{\partial v}{\partial n} - v \frac{\partial u}{\partial n} \right) dS \; , \tag{3}$$

where $\mathbf{A}$ is a continuously differentiable vector field and the functions u and v have partial derivatives of the second order which are continuous in the bounded region R; S is the boundary of R, while $\mathbf{n}$ stands for the unit normal outward to S. The surface S is a smooth (or regular) surface as defined by Kellog [9]. The differential operator ∇^2 is the Laplacian, which in Cartesian coordinates x_1, x_2, x_3 has the form

$$\nabla^2 = \frac{\partial^2}{\partial x_1^2} + \frac{\partial^2}{\partial x_2^2} + \frac{\partial^2}{\partial x_3^2} \; . \tag{4}$$

Any solution u of the equation $\nabla^2 u = 0$ is called a harmonic function.

To keep the chapter to a manageable size, we shall confine ourselves mainly to three-dimensional problems. The results can be readily extended to two-dimensional problems once the technique is grasped. Boldface letters such as $\mathbf{x}$ shall signify the triplet (x_1, x_2, x_3). The quantities R_i and R_e will stand for the regions interior and exterior to S, respectively. Furthermore, we shall not follow the common practice of writing dS_x or dS_ξ to signify that integration is with respect to the variable $\mathbf{x}$ or ξ. We shall merely write dS and it shall be clear from the context as to what the variable of integration is.

We shall be interested in giving the integral-equation formulation mainly to the differential equations $\nabla^2 u = 0$, the Laplace equation;

$\nabla^2 u = -4\pi\rho$, the Poisson equation; and $(\nabla^2 + k^2)u = 0$, the Helmholtz (or the steady-state wave) equation. Here, $\rho(\mathbf{x})$ is a given function of position and k is a given number.

6.2. INTEGRAL REPRESENTATION FORMULAS FOR THE SOLUTIONS OF THE LAPLACE AND POISSON EQUATIONS

Our starting point is the fundamental solution (or free-space solution) $E(\mathbf{x}; \xi)$ which satisfies

$$-\nabla^2 E = \delta(\mathbf{x} - \xi) \tag{1}$$

and vanishes at infinity. This function can be interpreted as the electrostatic potential at an arbitrary field point $\mathbf{x}$ due to a unit charge at the source point ξ. Such a potential is given as

$$E(\mathbf{x}; \xi) = 1/4\pi r = 1/4\pi |\mathbf{x} - \xi| . \tag{2}$$

For the two-dimensional case, the corresponding formula is $(1/2\pi)\log(1/r) = (1/2\pi)\log(1/|\mathbf{x} - \xi|)$, where $\mathbf{x} = (x_1, x_2)$ and $\xi = (\xi_1, \xi_2)$.

The fundamental solution can be used to obtain the solution of the Poisson equation

$$-\nabla^2 u = 4\pi\rho . \tag{3}$$

Indeed, multiply (1) by $u(x)$, (3) by $E(\mathbf{x}; \xi)$, subtract, integrate over the region R_i, and use Green's second identity as well as the sifting property of the delta function. The result is (after relabeling $\mathbf{x}$ and ξ)

$$u(\mathbf{x}) = \int_R \frac{\rho}{r} dV - \frac{1}{4\pi} \int_S u \frac{\partial}{\partial n}\left(\frac{1}{r}\right) dS + \frac{1}{4\pi} \int_S \frac{1}{r}\frac{\partial u}{\partial n} dS . \tag{4}$$

Suppose that from some previous considerations we know the values of ρ, u, and $\partial u/\partial n$ which appear in the formula (4):

$$u|_s = \tau , \qquad \partial u/\partial n|_s = \sigma ; \tag{5}$$

then this formula becomes

$$u(P) = \int_R \frac{\rho}{r} dV + \frac{1}{4\pi} \int_S \tau(Q)\frac{\partial}{\partial n}\left(\frac{1}{r}\right) dS - \frac{1}{4\pi} \int_S \frac{\sigma(Q)}{r} dS , \tag{6}$$

where P is the field point $\mathbf{x}$ and Q is a point ξ on S. The formulas (4) and (6) lead to many interesting properties of the harmonic functions. For details, the reader should refer to the appropriate texts $[4, 9, 19]$. We shall merely note the properties of the three integrals which occur on the right side of (6).

The Newtonian, Single-Layer, and Double-Layer Potentials

The integral $\int_R (\rho/r)\, dV$ is called the volume potential or Newtonian potential with volume density ρ. Similarly, the integral $\int_S (\sigma/r)\, dS$ is the simple or single-layer potential with charge (or source) density σ, while the integral $\int_S \tau(\partial/\partial n)(1/r)\, dS$ is the double-layer potential with dipole density τ. These integrals arise in all the fields of potential theory. However, we have used the language of electrostatics, as it interprets them nicely. These potentials have the following properties.

For the Newtonian potential $u = \int_R (\rho/r)\, dV$, we have the following:

(1) $\nabla^2 u = 0$, for points P in R_e.
(2) For points P within R_i, the integral is improper but it converges and admits two differentials under the integral sign if the function ρ is sufficiently smooth; the result is $\nabla^2 u = -4\pi\rho(P)$.

The single-layer potential $u = \int_S (\sigma/r)\, dS$ has the following properties:

(1) $\nabla^2 u = 0$, outside S.
(2) The integral becomes improper at the surface S but converges uniformly if S is regular. Moreover, this integral remains continuous as we pass through S.
(3) Consider the derivative of u taken in the direction of a line normal to the surface S in the outward direction from S. Then,

$$\left.\frac{\partial u}{\partial n}\right|_{P_+} = -2\pi\sigma(P) + \int_S \sigma(Q)\frac{\cos(\xi-\mathbf{x},\mathbf{n})}{|\mathbf{x}-\xi|^2}\, dS \tag{7}$$

and

$$\left.\frac{\partial u}{\partial n}\right|_{P_-} = 2\pi\sigma(P) + \int_S \sigma(Q)\frac{\cos(\xi-\mathbf{x},\mathbf{n})}{|\mathbf{x}-\xi|^2}\, dS , \tag{8}$$

where P_+ and P_- signify that we approach S from R_i and R_e, respectively, and where both $\mathbf{x}$ and $\boldsymbol{\xi}$ are on S. From (7) and (8), we obtain the jump of the normal derivative of u across S:

$$\sigma = \frac{1}{4\pi}\left(\frac{\partial u}{\partial n}\bigg|_{P_-} - \frac{\partial u}{\partial n}\bigg|_{P_+}\right). \tag{9}$$

Similarly, the double-layer potential $u = \int_S \tau (\partial/\partial n)(1/r)\, dS$, has the following properties:

(1) $\nabla^2 u = 0$ outside of S.

(2)   The integral becomes improper at the surface but it converges if the surface S is regular.

(3)   The integral undergoes a discontinuity when passing through S such that

$$u\big|_{P_+} = 2\pi\tau(P) + \int_S \tau(Q)\frac{\cos(\mathbf{x}-\boldsymbol{\xi}, \mathbf{n})}{|\mathbf{x}-\boldsymbol{\xi}|^2}\, dS \tag{10}$$

and

$$u\big|_{P_-} = -2\pi\tau(P) + \int_S \tau(Q)\frac{\cos(\mathbf{x}-\boldsymbol{\xi}, \mathbf{n})}{|\mathbf{x}-\boldsymbol{\xi}|^2}\, dS \tag{11}$$

in the notation of the relations (7) and (8). Hence, the jump of u across S is

$$\tau = (1/4\pi)\left[u\big|_{P_+} - u\big|_{P_-}\right]. \tag{12}$$

(4)   The normal derivative remains continuous as S is crossed. The reader who is interested in the proof should look up the elegant proof given by Stakgold [19].

Interior and Exterior Dirichlet Problems

For the solution of a boundary value problem for an elliptic equation, we cannot prescribe u and $\partial u/\partial n$ arbitrarily on S. Therefore, equation (6) does not allow us to construct a solution for equation (3) such that u shall itself have arbitrary values on S and also arbitrary values for its normal derivative there. As such, there are two kinds of boundary value problems for elliptic equations. For one kind, we have the value of the solution prescribed on S—the so-called Dirichlet problem. For the second kind, the value of the normal derivative of the solution is prescribed on S—the Neumann problem. We discuss the Dirichlet problem first.

To fix the ideas, let us discuss the Dirichlet problem for the region exterior to the unit sphere. In order to get a unique solution, it is necessary to impose some sort of boundary condition at infinity along with the boundary value on the surface S of the sphere. Indeed, the functions $u_1(\mathbf{x}) = 1$ and $u_2(\mathbf{x}) = 1/r$ are both harmonic in the region R_e and assume the same value 1 on S. But if we require that the solution vanishes at infinity, then u_2 is the required solution. As a matter of fact, it is an important result in the potential theory [9, 19] that, when one solves the Dirichlet problem for the exterior of the unit sphere (by expansions in spherical harmonics) such that the potential vanishes at infinity, then one finds that the behavior of the solution is

$$u\big|_\infty = O\!\left(\frac{1}{r}\right), \qquad \frac{\partial u}{\partial r}\bigg|_\infty = O\!\left(\frac{1}{r^2}\right). \tag{13}$$

From these considerations and from the value of the fundamental solution, it is traditionally proved that

$$\lim_{r\to\infty} \int_S \left(E\frac{\partial u}{\partial r} - u\frac{\partial E}{\partial r}\right) dS = 0 \tag{14}$$

on the surface of the sphere of radius r.

We can now define and analyze the exterior and interior Dirichlet problems for an arbitrary surface S as follows.

Definition. The exterior Dirichlet problem is the boundary value problem

$$\nabla^2 u_e = 0, \qquad \mathbf{x} \in R_e; \qquad u_e\big|_S = f,$$

$$u_e\big|_\infty = O\!\left(\frac{1}{r}\right), \qquad \frac{\partial u_e}{\partial r}\bigg|_\infty = O\!\left(\frac{1}{r^2}\right), \tag{15}$$

where $f(\mathbf{x})$ is a given continuous function on S.

Definition. The interior Dirichlet problem is the boundary value problem

$$\nabla^2 u_i = 0 , \qquad \mathbf{x} \in R_i , \qquad u_i|_S = f . \tag{16}$$

Suppose that we are required to find the solution of the interior Dirichlet problem. We assume that such a solution u is the potential of a double layer with density τ (which is as yet unknown):

$$u_i(\mathbf{x}) = \int_S \frac{\tau(\xi)\cos(\mathbf{x}-\xi, \mathbf{n})}{r^2} \, dS . \tag{17}$$

For u_i to satisfy the boundary condition $(16)_2$ from within S, we appeal to the relation (11) and get the Fredholm integral equation of the second kind for $\tau(P)$:

$$\tau(P) = -(1/2\pi)f(P) + \int_S K(P, Q)\tau(Q) \, dS , \tag{18}$$

where the kernel $K(P, Q)$ is

$$K(P, Q) = \left[\cos(\mathbf{x}-\xi, \mathbf{n})\right]/2\pi \, |\mathbf{x}-\xi|^2 , \tag{19}$$

and $P \,(=\mathbf{x})$ and $Q \,(=\xi)$ are both on S. We solve the integral equation (18) for τ, substitute this solution in (17), and obtain the required solution of the boundary value problem (16).

In exactly the same way, the Dirichlet problem for an external domain bounded internally by S can be reduced to the solution of a Fredholm integral equation of the second kind.

We can present an integral-equation formulation of the exterior and interior boundary value problems (15) and (16) in a composite medium when $f(\mathbf{x})$ is the same function in both these problems. Recall that the fundamental solution $E(\mathbf{x}; \xi)$ satisfies the relation

$$-\nabla^2 E = \delta(\mathbf{x}-\xi) , \qquad \text{for all } \mathbf{x} \text{ and } \xi . \tag{20}$$

Multiply (16) by E, (20) by u_i, add, integrate and apply Green's second identity. This results in

$$\int_S \left(E\frac{\partial u_i}{\partial n} - u_i\frac{\partial E}{\partial n} \right) dS = \begin{cases} u_i(\xi) , & \xi \in R_i , \\ 0 , & \xi \in R_e , \end{cases} \tag{21}$$

where $\mathbf{n}$ is the outward normal to R_i on S.

The corresponding result for the exterior region is obtained by multiplying (15) by E, (20) by u_e, adding, integrating over the region bounded internally by S and externally by a sphere S_r, and applying Green's second identity. The contribution from S_r vanishes as $r \to \infty$ in view of the boundary condition at infinity, and we finally have

$$\int_S \left(-E \frac{\partial u_e}{\partial n} + u_e \frac{\partial E}{\partial n} \right) dS = \begin{cases} 0, & \xi \in R_i, \\ u_e(\xi), & \xi \in R_e, \end{cases} \tag{22}$$

where we have used the fact that the outward normal to R_e on S is in the $-\mathbf{n}$ direction.

The next step is to add (21) and (22) and observe that both u_i and u_e take the same value f as we approach the surface. Thus, we obtain

$$\int_S E(\mathbf{x}; \xi) \left(\frac{\partial u_i}{\partial n} - \frac{\partial u_e}{\partial n} \right) dS = \begin{cases} u_i(\xi), & \xi \in R_i, \\ u_e(\xi), & \xi \in R_e, \end{cases} \tag{23}$$

and $\mathbf{x} \in S$. Let us make use of the relations (2) and (9), and relabel $\mathbf{x}$ and ξ; thereby, we end up with the relation

$$\int_S \frac{\sigma(\xi)}{|\mathbf{x} - \xi|} dS = \begin{cases} u_i(\mathbf{x}), & \mathbf{x} \in R_i, \\ u_e(\mathbf{x}), & \mathbf{x} \in R_e, \end{cases} \tag{24}$$

that is, a single-layer potential with unknown charge density σ. Finally, using the boundary condition

$$u_i|_S = u_e|_S = f,$$

in (24), we obtain the Fredholm integral equation of the first kind

$$f(\mathbf{x}) = \int_S [\sigma(\xi)/|\mathbf{x} - \xi|] \, dS, \tag{25}$$

with both $\mathbf{x}$ and ξ on S.

Interior and Exterior Neumann Problems

In this case we are required to find the solution of Laplace or Poisson equation when the normal derivative is prescribed.

Definition. The exterior Neumann problem is the boundary value problem

$$\nabla^2 u_e = 0 , \qquad \mathbf{x} \in R_e , \qquad \left.\frac{\partial u_e}{\partial n}\right|_s = f , \qquad u_e|_\infty = 0 . \qquad (26)$$

Definition. The interior Neumann problem is the boundary value problem

$$\nabla^2 u_i = 0 , \qquad \mathbf{x} \in R_i , \qquad \left.\frac{\partial u_i}{\partial n}\right|_s = f . \qquad (27)$$

For a Neumann problem, the prescribed function $f(\mathbf{x})$ satisfies the consistency condition

$$\int_S f(\xi) \, dS = 0 , \qquad \xi \in S , \qquad (28)$$

which follows by integrating the identity

$$\int_{R_i} (\nabla^2 u_i) \, dV = 0 ,$$

and using the divergence theorem.

The exterior and interior Neumann problems can be reduced to integral equations in a manner similar to the one explained for the corresponding Dirichlet problem. Indeed, we seek a solution of the interior Neumann problem in the form of the potential of a simple layer

$$u_i = \int_S [\sigma(Q)/r] \, dS , \qquad (29)$$

which is a harmonic function in R_i. It will be a solution of (27) if the density σ is so chosen that

$$\left.\frac{\partial u_i}{\partial n}\right|_{P_-} = f(P) , \qquad P \in S . \qquad (30)$$

Appealing to the relation (8), we have

$$f(P) = \left.\frac{\partial u_i}{\partial n}\right|_{P_-} = 2\pi\sigma(P) + \int_S \sigma(Q) \frac{\cos(\xi - \mathbf{x}, \mathbf{n})}{r^2} \, dS . \qquad (31)$$

Thus, $\sigma(P)$ is a solution of the Fredholm integral equation of the second kind

$$\sigma(P) = \frac{1}{2\pi} f(P) - \int_S \sigma(Q) \frac{\cos(\xi - \mathbf{x}, \mathbf{n})}{2\pi r^2} \, dS . \tag{32}$$

The solution of the exterior Neumann problem also leads to a similar integral equation. Furthermore, we can give the integral-equation formulation of the problems (26) and (27) in a composite medium when f is the same function in both these problems. Proceeding as we did for the corresponding Dirichlet problem, we obtain a Fredholm integral equation of the first kind. Instead of a single-layer potential, we now get a double-layer potential. The details are left to the reader.

Let us observe in passing that the solution of (27) is not unique, since an arbitrary constant can be added to a solution and the resulting function will satisfy (27).

6.3. EXAMPLES

Example 1. *Electrostatic potential due to a thin circular disk.* Let us take S to be a circular disk of radius a on which the potential V is prescribed. Let us choose cylindrical polar coordinates (ρ, φ, z) such that the origin is on the center of the disk with the z axis normal to the plane of the disk. Thus, the disk occupies the region $z = 0$, $0 \leqslant \rho \leqslant a$, for all φ. There is no loss of generality in taking the potential V on the disk as $f^{(n)}(\rho) \cos n\varphi$, where n is an arbitrary integer, because we can use the Fourier superposition principle. The charge density σ will also then have the form $\sigma^{(n)}(\rho) \cos n\varphi$. From (6.2.25), we have

$$f^{(n)}(\rho) \cos n\varphi = \int_{\text{disk}} [\sigma(\xi)/|\mathbf{x} - \xi|] \, dS , \tag{1}$$

where $\mathbf{x} = (\rho, \varphi, 0)$ and $\xi = (t, \varphi_1, 0)$. Or

$$f^{(n)}(\rho) \cos n\varphi = \int_0^a \int_0^{2\pi} \frac{t\sigma^{(n)}(t)(\cos n\varphi_1) \, d\varphi_1 \, dt}{[\rho^2 + t^2 - 2\rho t \cos(\varphi - \varphi_1)]^{\frac{1}{2}}} . \tag{2}$$

But, by setting $\varphi_1 - \varphi = \psi$, we find that

$$\int_0^{2\pi} \frac{\cos n\varphi_1 \, d\varphi_1}{[\rho^2 + t^2 - 2\rho t \cos(\varphi - \varphi_1)]^{1/2}} = \int_{-\varphi}^{2\pi-\varphi} \frac{\cos n(\varphi + \psi) \, d\psi}{[\rho^2 + t^2 - 2\rho t \cos(\varphi - \varphi_1)]^{1/2}}$$

$$= \int_{-\varphi}^{2\pi-\varphi} \frac{(\cos n\varphi)(\cos n\psi) \, d\psi}{[\rho^2 + t^2 - 2\rho t \cos \psi]^{1/2}}$$

$$= (\cos n\varphi) \left[\int_{-\varphi}^{0} + \int_{0}^{2\pi} + \int_{2\pi}^{2\pi-\varphi} \frac{\cos n\psi \, d\psi}{[\rho^2 + t^2 - 2\rho t \cos \psi]^{1/2}} \right]$$

$$= (\cos n\varphi) \int_0^{2\pi} \frac{\cos n\psi \, d\psi}{[\rho^2 + t^2 - 2\rho t \cos \psi]^{1/2}} , \tag{3}$$

where in the first integral we have put $\psi' = \psi + 2\pi$.

From (2) and (3), there follows the equation

$$f^{(n)}(\rho) = \int_0^a \int_0^{2\pi} \frac{t\sigma^{(n)}(t) \cos n\psi \, d\psi \, dt}{[\rho^2 + t^2 - 2\rho t \cos \psi]^{1/2}} . \tag{4}$$

Finally, we use in (4) the expansion formula

$$[\rho^2 + t^2 - 2\rho t \cos \psi]^{-1/2} = \sum_{r=0}^{\infty} \int_0^{\infty} (2 - \delta_{0r})(\cos r\psi) J_r(p\rho) J_r(pt) \, dp , \tag{5}$$

where δ_{0r} is the Kronecker delta, and use the orthogonality of the cosine function. The result is the Fredholm integral equation of the first kind

$$f^{(n)}(\rho) = \int_0^a t\sigma^{(n)}(t) K_0(t, \rho) \, dt , \tag{6}$$

where the kernel $K_0(t, \rho)$ is

$$K_0(t, \rho) = 2\pi \int_0^{\infty} J_n(p\rho) J_n(pt) \, dp . \tag{7}$$

For an annular disk of inner radius b and outer radius a, the formula that corresponds to (6) is

$$f^n(\rho) = \int_b^a t\sigma^n(t) K_0(t,\rho) \, dt \, . \tag{8}$$

Example 2. Solve the integral equation (6.2.25) when S is a unit sphere and $f = \sin\theta\cos\varphi$; that is, solve the integral equation

$$\sin\theta\cos\varphi = \int_0^{2\pi} d\varphi_1 \int_0^\pi \frac{(\sin\theta_1)\,\sigma(\theta_1,\varphi_1)\,d\theta_1}{|\mathbf{x}-\boldsymbol{\xi}|} \, . \tag{9}$$

Here, we use the expansion formula

$$\frac{1}{|\mathbf{x}-\boldsymbol{\xi}|} = \sum_{n=0}^\infty N_{0,n} \sum_{m=-n}^n \frac{Y_n^m(\theta,\varphi)\,Y_n^{*m}(\theta_1,\varphi_1)}{N_{m,n}} \, , \tag{10}$$

where $Y_n^m(\theta,\varphi)$ are the spherical harmonics and

$$N_{m,n} = \int_0^{2\pi} d\varphi \int_0^\pi (\sin\theta)\,|Y_n^m(\theta,\varphi)|^2 \, d\theta$$

$$= \frac{4\pi}{2n+1} \frac{(n+|m|)!}{(n-|m|)!} \, . \tag{11}$$

Furthermore, we set

$$\sigma(\theta_1,\varphi_1) = \sum_{m=0}^\infty \sum_{m=-n}^n \sigma_{m,n} Y_n^m(\theta_1,\varphi_1) \, , \tag{12}$$

and note that $\sin\theta\cos\varphi = \frac{1}{2}[Y_1^1(\theta,\varphi) + Y_1^{-1}(\theta,\varphi)]$. Putting (10)–(12) in (9) and using the orthogonality properties of the spherical harmonics, we obtain

$$\sigma_{1,1} = \sigma_{1,-1} = 3/8\pi \, ,$$

and $\sigma_{m,n} \equiv 0$ for all other m and n. Thus, from (12), it follows that

$$\sigma(\theta,\varphi) = (3/4\pi) P_1^1(\cos\theta)\cos\varphi \, , \tag{13}$$

where P_1^1 is an associated Legendre function.

6.4. GREEN'S FUNCTION APPROACH

The Green's function is an auxiliary function which plays the same crucial role in the integral-equation formulation of partial differential equations as it plays in the case of ordinary differential equations. This function depends on the form of the differential equation, the boundary condition, and the region. For instance, the Green's function $G(\mathbf{x}; \xi)$ for the Laplace equation in an open, bounded region R in three-dimensional space with boundary S is the solution of the boundary value problem

$$-\nabla^2 G = \delta(\mathbf{x} - \xi), \qquad G|_S = 0, \tag{1}$$

where $\mathbf{x}$ and ξ are in R. In the language of electrostatics, the function G is the electrostatic potential due to a unit charge at ξ when the surface S is a grounded metallic shell. As such, G is the sum of the potential of the unit source at ξ in free space and the potential due to the charge induced on S:

$$G(\mathbf{x}; \xi) = (1/4\pi |\mathbf{x} - \xi|) + v(\mathbf{x}; \xi), \tag{2}$$

where v is a harmonic function which satisfies the boundary value problem

$$\nabla^2 v = 0, \qquad \mathbf{x} \in R, \qquad v|_S = -E. \tag{3}$$

Let us show that the Green's function is symmetric. When $G(\mathbf{x}; \xi)$ and $G(\mathbf{x}; \eta)$ are the Green's functions for the region R corresponding to the sources at ξ and η, we have the relations

$$-\nabla^2 G(\mathbf{x}; \xi) = \delta(\mathbf{x} - \xi), \qquad G|_S = 0, \tag{4}$$

$$-\nabla^2 G(\mathbf{x}; \eta) = \delta(\mathbf{x} - \eta), \qquad G|_S = 0. \tag{5}$$

The result of the routine steps of multiplying (4) by $G(\mathbf{x}; \eta)$, (5) by $G(\mathbf{x}; \xi)$, subtracting, integrating, and applying Green's second identity is

$$\int_S \left[G(\mathbf{x}; \xi) \frac{\partial G(\mathbf{x}; \eta)}{\partial n} - G(\mathbf{x}; \eta) \frac{\partial G(\mathbf{x}; \xi)}{\partial n} \right] dS = G(\xi; \eta) - G(\eta; \xi). \tag{6}$$

The symmetry of the Green's function follows by applying the boundary conditions in (6).

Note also that the fundamental solution $E(\mathbf{x};\xi)$ is the free-space Green's function.

Solution of the Dirichlet Problem

We are now ready to give an integral-equation formulation to the boundary value problem

$$-\nabla^2 u(\mathbf{x}) = 4\pi\rho(\mathbf{x}), \qquad \mathbf{x} \in R, \qquad u|_S = f, \tag{7}$$

in terms of the Green's function. For this purpose, multiply (1) by u and (7) by G, subtract, integrate, use Green's second identity, and get

$$u(\xi) = 4\pi \int_R G(\mathbf{x};\xi)\,\rho(\mathbf{x})\,dV - \int_S f(\mathbf{x})\,[\partial G(\mathbf{x};\xi)/\partial n]\,dS. \tag{8}$$

By interchanging $\mathbf{x}$ and ξ and using the symmetry of the Green's function, we find the representation formula

$$u(\mathbf{x}) = 4\pi \int_R G(\mathbf{x};\xi)\,\rho(\xi)\,dV - \int_S [\partial G(\mathbf{x};\xi)/\partial n]f(\xi)\,dS. \tag{9}$$

For the particular case $\rho \equiv 0$, the formula (9) becomes

$$u(\mathbf{x}) = -\int_S f(\xi)\,[\partial G(\mathbf{x};\xi)/\partial n]\,dS. \tag{10}$$

When $f = 1$ on S, then the solution u of the Laplace equation is clearly $u = 1$ for the interior Dirichlet problem. Thus, (10) yields an interesting relation,

$$-\int_S [\partial G(\mathbf{x};\xi)]/\partial n]\,dS = 1, \qquad \mathbf{x} \in R. \tag{11}$$

Example. *Poisson integral formula.* The Green's function for the Laplace equation, when the surface S is a sphere, can be found by various methods. The easiest method is to express it as source and image point combination. Let the radius of the sphere be a (see Figure 6.1).

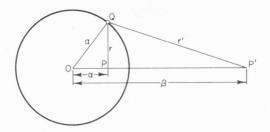

Figure 6.1

For any point P ($=\mathbf{x}$) with radial distance α within the sphere, we have an inverse point P' ($=\mathbf{x}'$) on the same radial line at a radial distance β outside the sphere, such that $\alpha\beta = a^2$. If Q ($=\xi$) is any point on S, then the triangles OQP and OQP' are easily seen to be similar. Therefore, $r'/r = a/\alpha$, or

$$1/r = a/\alpha r' . \tag{12}$$

Examining the relations (1) and (2), we readily find the value of the Green's function to be

$$G(P, Q) = \frac{1}{4\pi}\left(\frac{1}{r} - \frac{a}{\alpha}\frac{1}{r'}\right). \tag{13}$$

Having found the Green's function, we can solve the interior Dirichlet problem for the sphere:

$$\nabla^2 u = 0, \quad r < a; \quad u = f(\theta, \varphi) \quad \text{on} \quad r = a. \tag{14}$$

To use the formula (10) we need the value of $\partial G/\partial n$. This is obtained if we observe from the figure that

$$\alpha^2 = a^2 + r^2 - 2ar\cos(\mathbf{x}-\xi, \mathbf{n}) ;$$
$$\beta^2 = a^2 + r'^2 - 2ar'\cos(\mathbf{x}'-\xi, \mathbf{n}) . \tag{15}$$

Thus,

$$\frac{\partial G}{\partial n} = \frac{1}{4\pi}\left(-\frac{1}{r^2}\frac{\partial r}{\partial n} + \frac{a}{\alpha}\frac{1}{r'^2}\frac{\partial r'}{\partial n}\right) = \frac{1}{4\pi}\left(\frac{\cos(\mathbf{x}-\boldsymbol{\xi},\mathbf{n})}{r^2} - \frac{a}{\alpha}\frac{\cos(\mathbf{x'}-\boldsymbol{\xi},\mathbf{n})}{r'^2}\right)$$

$$= \frac{1}{4\pi}\left(\frac{\alpha^2-a^2-r^2}{2ar^3} - \frac{a}{\alpha}\frac{\beta^2-a^2-r'^2t}{2ar'^3}\right) = -\frac{a^2-\alpha^2}{4\pi ar^3}, \qquad (16)$$

where we have used the relations $r'/r = a/\alpha$ and $\alpha\beta = a^2$.

Substituting (16) in (10), we finally have

$$u(P) = \frac{a(a^2-\alpha^2)}{4\pi}\int\limits_0^\pi\int\limits_0^{2\pi}\frac{f(\theta',\varphi')(\sin\theta')\,d\theta'\,d\varphi'}{(\alpha^2+a^2-2a\alpha\cos(\mathbf{x},\boldsymbol{\xi})^{3/2}}. \qquad (17)$$

Neumann Problem

By defining the Green's function $G(\mathbf{x};\boldsymbol{\xi})$ by the boundary value problem

$$-\nabla^2 G(\mathbf{x};\boldsymbol{\xi}) = \delta(\mathbf{x}-\boldsymbol{\xi}), \qquad \partial G/\partial n\big|_S = 0, \qquad (18)$$

we can extend the above analysis to the Neumann problem. Indeed, the integral equation that corresponds to (10) for the Neumann problem

$$-\nabla^2 u = 0, \qquad \mathbf{x}\in R, \qquad \partial u/\partial n = f \qquad (19)$$

is

$$u(\mathbf{x}) = \int_S G(\mathbf{x};\boldsymbol{\xi})f(\boldsymbol{\xi})\,dS. \qquad (20)$$

Of course, the function f must satisfy the consistency condition (6.2.28).

Finally, let us consider the interior and exterior Dirichlet problems for a body S enclosed within a surface Σ:

$$\nabla^2 u_i = 0, \qquad \mathbf{x}\in R_i, \qquad u_i\big|_S = f, \qquad (21)$$

$$\nabla^2 u_e = 0, \qquad \mathbf{x}\in R_e, \qquad u_e\big|_S = f, \qquad u_e\big|_\Sigma = 0. \qquad (22)$$

The Green's function $G(\mathbf{x};\boldsymbol{\xi})$ satisfies the auxiliary problem (we absorb the factor 4π in G):

$$-\nabla^2 G = 4\pi\delta(\mathbf{x}-\boldsymbol{\xi}), \qquad G\big|_\Sigma = 0. \qquad (23)$$

We now follow the same steps as we did in deriving the formula (6.2.25). Thus, from the relations (21)–(23), we obtain

$$f(\mathbf{x}) = \int_S G(\mathbf{x};\xi)\sigma(\xi)\,dS\,, \tag{24}$$

which reduces to (6.2.25) for an unbounded medium.

6.5. EXAMPLES

Example 1. *Electrostatic potential problem of a conducting disk bounded by two parallel planes.* This problem is an extension of the problem considered in Example 1 of Section 6.3. We follow that notation and assume that the parallel planes are $z = b$ and $z = -c$ $(b, c > 0)$. The boundary value problem becomes

$$\nabla^2 V(\rho, \varphi, z) = 0 \quad \text{in} \quad D\,, \tag{1}$$

$$V(\rho, \varphi, 0) = f^{(n)}(\rho)\cos n\varphi\,, \qquad 0 \leqslant \rho \leqslant a\,, \tag{2}$$

$$V(\rho, \varphi, z) = 0\,, \qquad z = b\,, \qquad z = -c\,, \tag{3}$$

where D is the region between the disk and the parallel planes.

The Green's function G corresponding to this problem satisfies the auxiliary system

$$-\nabla^2 G(\mathbf{x};\xi) = 4\pi\delta(\mathbf{x}-\xi)\,, \qquad G = 0 \quad \text{on} \quad z = b\,, \quad z = -c\,. \tag{4}$$

This function is found easily by the method of images. Indeed, for a positive unit charge at the source point $\xi = (t, \varphi_1, z_1)$, the image system consists of positive unit charge at the points (see Figure 6.2)

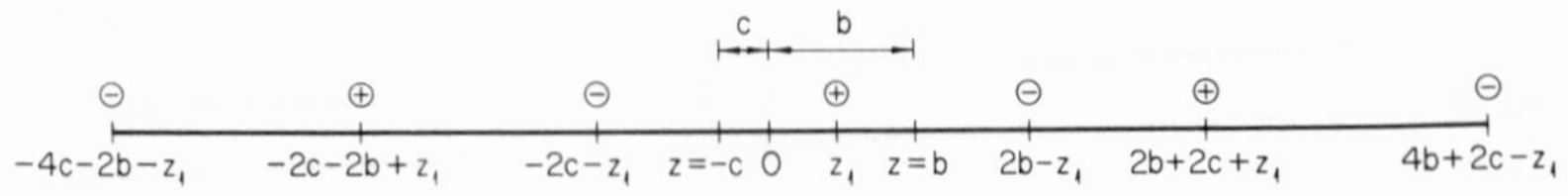

Figure 6.2

$$\xi_n^+ = [t, \varphi_1, 2n(b+c) + z_1], \qquad n = \pm 1, \pm 2, \dots, \tag{5}$$

and negative unit charge at points

$$\xi_n^- = [t, \varphi_1, 2n(b+c) - 2c - z_1], \qquad n = 0, \pm 1, \pm 2, \dots. \tag{6}$$

The value of the Green's function therefore is

$$G(\mathbf{x};\xi) = \frac{1}{|\mathbf{x}-\xi|} + \sum_{n=1}^{\infty} \frac{1}{|\mathbf{x}-\xi_n^+|} + \sum_{n=-1}^{-\infty} \frac{1}{|\mathbf{x}-\xi_n^+|} - \sum_{n=-\infty}^{\infty} \frac{1}{|\mathbf{x}-\xi_n^-|}. \tag{7}$$

The next step is to use the identity

$$1/|\mathbf{x}-\xi| = \int_0^{\infty} J_0(p\varpi)(\exp -p|z-z_1|)\, dp, \tag{8}$$

where $\varpi = [\rho^2 + t^2 - 2\rho t \cos(\varphi - \varphi_1)]^{1/2}$. Then, the relation (7) takes the form

$$\begin{aligned} G(\mathbf{x};\xi) = (1/|\mathbf{x}-\xi|) + \int_0^{\infty} J_0(p\varpi) \Big[&\sum_{1}^{\infty} (\exp -p|z - 2n(b+c) - z_1|) \\ &+ \sum_{-1}^{-\infty} (\exp -p|z - 2n(b+c) - z_1|) \\ &- \sum_{0}^{\infty} (\exp -p|z - 2n(b+c) + 2c + z_1|) \\ &- \sum_{-1}^{-\infty} (\exp -p|z - 2n(b+c) + 2c + z_1|) \Big] dp. \end{aligned} \tag{9}$$

After summing the geometric series which occur in relation (9) and slight simplification, we obtain

$$\begin{aligned} G(\mathbf{x};\xi) = \frac{1}{|\mathbf{x}-\xi|} + \int_0^{\infty} J_0(p\varpi) \\ \times \frac{e^{-(c+z_1)p}[\sinh p(z-b)] - e^{-(n-z_1)p}[\sinh p(z+c)]}{\sinh p(b+c)}\, dp. \end{aligned} \tag{10}$$

Finally, the result of using the expansion of $J_0(p\varpi)$,

$$J_0(p\varpi) = \sum_{r=0}^{\infty} (2-\delta_{0r})[\cos r(\varphi-\varphi_1)]J_r(p\rho)J_r(pt), \qquad (11)$$

in (10) is

$$G(\mathbf{x};\boldsymbol{\xi}) = \frac{1}{|\mathbf{x}-\boldsymbol{\xi}|} + \sum_{r=0}^{\infty} (2-\delta_{0r})[\cos r(\varphi-\varphi_1)\,G^{(r)}(\rho,t,z,z_1), \qquad (12)$$

where

$$G^{(r)}(\rho,t,z,z_1) = \int_0^{\infty} \frac{e^{-p(c+z_1)}[\sinh p(z-b)] - e^{-p(b-z_1)}[\sinh p(z+c)]}{\sinh p(b+c)}$$

$$\times J_r(p\rho)J_r(pt)\,dp . \qquad (13)$$

To derive the integral equation, we multiply (1) by G and (4) by V, follow the routine procedure, and get

$$V(\rho,\varphi,z) = \frac{1}{4\pi} \int_{S^++S^-} \left(G\frac{\partial V}{\partial n} - V\frac{\partial G}{\partial n}\right) dS, \qquad (14)$$

where S^+ and S^- are the upper and lower parts of the disk, respectively. On the surfaces $S^{\pm}$, the value of the outward normal is $\mp\partial/\partial z_1$, respectively. Using this fact and the boundary conditions (2) in (14), we have

$$f^{(n)}(\rho)\cos n\varphi = \int_0^a\int_0^{2\pi} t\sigma^{(n)}(t)\,G(\rho,t,\varphi,\varphi_1,0,0)(\cos n\varphi_1)\,dp_1\,dt, \qquad (15)$$

where $\sigma^{(n)}(t)\cos n\varphi_1 = (1/4\pi)(\partial V/\partial z_{1+} - \partial V/\partial z_{1-})$. Setting $\varphi_1-\varphi=\psi$, and following the steps which led from (6.3.2) to (6.3.4), we obtain

$$f^{(n)}(\rho) = \int_0^a t\sigma^{(n)}(t)\,dt \int_0^{2\pi} \frac{\cos n\psi\,d\psi}{|\mathbf{x}-\boldsymbol{\xi}|_{z=z_1=0}}$$

$$+ \int_0^a t\sigma^{(n)}(t)[2\pi G^{(n)}(\rho,t,0,0)]\,dt, \qquad (16)$$

where we have substituted the value of G as in (12). This can be written as

$$f^n(\rho) = \int_0^a t\sigma^{(n)}(t) K_1(t,\rho) \, dt \,, \tag{17}$$

where

$$K_1(t,\rho) = \int_0^{2\pi} \frac{\cos n\psi \, d\psi}{[\rho^2 + t^2 - 2\rho t \cos \psi]^{\frac{1}{2}}} + 2\pi G^{(n)}(\rho, t, 0, 0) \,. \tag{18}$$

When $b \to \infty$, $c \to \infty$, we recover the formula (6.3.4).

For an annular disk of inner radius b and outer radius a, the formula which corresponds to (17) is

$$f^{(n)}(\rho) = \int_b^a t\sigma^{(n)}(t) K_1(t,\rho) \, dt \tag{19}$$

with the same kernel as in (18).

Example 2. *Electrostatic potential problem of an axially symmetric conductor placed symmetrically inside a cylinder of radius* b. We again take cylindrical polar coordinates (ρ, φ, z) with origin at the center of the conductor and z axis along its axis of symmetry (which is also the axis of the cylinder). For simplicity, we take V on the surface S of the conductor to be unity. Then, from the relation (6.4.24), we have the Fredholm integral equation of the first kind

$$1 = \int_S G(\mathbf{x};\boldsymbol{\xi})\sigma(\boldsymbol{\xi}) \, dS \,, \qquad \mathbf{x}, \boldsymbol{\xi} \in S \,, \tag{20}$$

where G satisfies the system

$$-\nabla^2 G = 4\pi\delta(\mathbf{x}-\boldsymbol{\xi}) \,, \qquad G = 0 \quad \text{on} \quad \rho = b \,. \tag{21}$$

In terms of cylindrical polar coordinates, the differential equation for G becomes

$$\frac{1}{\rho}\left(\rho \frac{\partial G}{\partial \rho}\right) + \frac{1}{\rho^2}\frac{\partial^2 G}{\partial \varphi^2} + \frac{\partial^2 G}{\partial z^2} = -\frac{4\pi}{\rho}\delta(\rho-t)\delta(\psi)\delta(z-z_1) \,, \tag{22}$$

where $\varphi_1 - \varphi = \psi$. From the definition (6.4.2) of the Green's function, we know that

$$G_1(\mathbf{x};\boldsymbol{\xi}) = G(\mathbf{x};\boldsymbol{\xi}) - (1/|\mathbf{x}-\boldsymbol{\xi}|) \tag{23}$$

is finite in the limit as $\mathbf{x} \to \boldsymbol{\xi}$. We can calculate the solution of (22) by taking the Fourier series expansion

$$G(\mathbf{x};\boldsymbol{\xi}) = \sum_{r=1}^{\infty} (2-\delta_{0r})(\cos r\psi) g^{(r)}(\rho,t,z,z_1) , \qquad (24)$$

where

$$g^{(r)} = (1/2\pi) \int_0^{2\pi} G(\mathbf{x};\boldsymbol{\xi})(\cos r\psi) \, d\psi .$$

Multiply the differential equation (22) by $(1/2\pi)\cos r\psi$ and integrate with respect to ψ from 0 to 2π. The result is

$$\frac{1}{\rho}\frac{\partial}{\partial\rho}\left(\rho\frac{\partial}{\partial\rho} g^{(r)}\right) - \frac{r^2}{\rho^2} g^{(r)} + \frac{\partial^2 g^{(r)}}{\partial z^2} = -\frac{2}{\rho}\delta(\rho-\rho_1)\delta(z-z_1) . \qquad (25)$$

Next, we take the Fourier transform of equation (25) by setting

$$T(g^{(r)}) = (2\pi)^{-\frac{1}{2}} \int_{-\infty}^{\infty} e^{ipz} g^{(r)} \, dz ;$$

$$g^{(r)} = (2\pi)^{-\frac{1}{2}} \int_{-\infty}^{\infty} e^{-ipz} T(g^{(r)}) \, dp . \qquad (26)$$

The system (21) becomes

$$\rho^2 \frac{d^2}{d\rho^2} T(g^{(r)}) + \rho \frac{d}{d\rho} T(g^{(r)}) - (\rho^2 p^2 + r^2) T(g^{(r)})$$

$$= -\frac{\sqrt{2}}{\rho\sqrt{\pi}} e^{ipz_1} \delta(\rho-\rho_1) , \qquad T(g^{(r)}) = 0 , \qquad \rho = b . \qquad (27)$$

This boundary value problem can be easily solved by the method and notation of Chapter 5 and the solution so obtained is then inverted to yield

$$g^{(r)} = (1/\pi) \int_{-\infty}^{\infty} e^{ip(z_1-z)} \{K_r(p\rho_>) I_r(p\rho_<)$$

$$- [K_r(pb)/I_r(pb)] I_r(p\rho) I_r(pt)\} \, dp , \qquad (28)$$

where I_r and K_r are modified Bessel functions. Finally, from (24) and (28), we find the value of G:

$$G(\mathbf{x};\boldsymbol{\xi}) = (1/\pi) \sum_{r=0}^{\infty} (2-\delta_{0r})(\cos r\psi) \int_{-\infty}^{\infty} e^{ip(z_1-z)}$$

$$\times \{K_r(p\rho_>) I_r(p\rho_<) - [K_r(pb)/I_r(pb)] I_r(p\rho) I_r(pt)\} \, dp. \quad (29)$$

When $b \to \infty$, $G = 1/|\mathbf{x}-\boldsymbol{\xi}|$, and we have from (29),

$$1/|\mathbf{x}-\boldsymbol{\xi}| = (1/\pi) \sum_{r=0}^{\infty} (2-\delta_{0r})(\cos r\psi) \int_{-\infty}^{\infty} e^{ip(z_1-z)} K_r(p\rho_>) I_r(p\rho_<) \, dp. \quad (30)$$

Combining (29) and (30), we have

$$G(\mathbf{x};\boldsymbol{\xi}) = (1/|\mathbf{x}-\boldsymbol{\xi}|) + \sum_{r=0}^{\infty} (2-\delta_{0r})(\cos r\psi) G^{(r)}(\rho,t,z,z_1), \quad (31)$$

where

$$G^{(r)}(\rho,t,z,z_1) = -(1/\pi) \int_{-\infty}^{\infty} e^{-ip(z-z_1)} I_r(p\rho) I_r(pt) [K_r(pb)/I_r(pb)] \, dp$$

$$= -(1/\pi) \left[\int_{0}^{\infty} e^{-ip(z-z_1)} I_r(p\rho) I_r(pt) [K_r(pb)/I_r(pb)] \, dp \right.$$

$$\left. + \int_{-\infty}^{-0} e^{-ip(z-z_1)} I_r(p\rho) I_r(pt) [K_r(pb)/I_r(pb)] \, dp \right]. \quad (32)$$

Changing p to $-p$ in the second integral and observing that

$$I_r(-z) = (-1)^r I_r(z), \qquad K_r(-z) = (-1)^r K_r(z),$$

and hence $K_r(-pb)/I_r(-pb) = K_r(pb)/I_r(pb)$, we have from (32)

$$G^{(r)}(\rho,t,z,z_1) = -(2/\pi) \int_{0}^{\infty} \left[\frac{I_r(p\rho) I_r(pt) K_r(pb)}{I_r(pb)} \right] \cos[p(z-z_1)] \, dp. \quad (33)$$

So far, we have not used the fact that the conductor is axially symmetric. For an axially symmetric body, the Green's function is independent of φ. That leaves only one term in the series (31):

$$G(\mathbf{x};\boldsymbol{\xi}) = (1/|\mathbf{x}-\boldsymbol{\xi}|) + G^{(0)}(\rho,t,z,z_1). \quad (34)$$

Equation (34) is of the form (23). Substituting (34) in (20), we have the required integral equation. When $b \to \infty$, we recover equation (6.3.4) for $f^{(n)}(\rho) = 1$.

6.6. THE HELMHOLTZ EQUATION

The discussion of the previous two sections can be easily extended to the case of the Helmholtz equation

$$(\nabla^2 + \lambda) u = 0 . \tag{1}$$

The free-space Green's function or the fundamental solution $E(\mathbf{x};\boldsymbol{\xi})$ is the solution of the spherically symmetric differential equation

$$-\nabla^2 E - \lambda E = \delta(\mathbf{x} - \boldsymbol{\xi}) , \tag{2}$$

and which vanishes at infinity. Such a solution in three dimensions is

$$E(\mathbf{x};\boldsymbol{\xi}) = \frac{\exp(i|\mathbf{x}-\boldsymbol{\xi}|\sqrt{\lambda})}{4\pi|\mathbf{x}-\boldsymbol{\xi}|} = \frac{\exp(ir\sqrt{\lambda})}{4\pi r} . \tag{3}$$

(i) When λ is a complex number, then $\sqrt{\lambda}$ is selected to be that root of λ that has a positive imaginary part so that E vanishes exponentially at infinity.

(ii) When λ is real and positive, that is, $\lambda = \omega^2$, ω real, the solution

$$E(\mathbf{x};\boldsymbol{\xi}) = \frac{\exp i\omega|\mathbf{x}-\boldsymbol{\xi}|}{4\pi|\mathbf{x}-\boldsymbol{\xi}|} = \frac{\exp i\omega r}{4\pi r} \tag{4}$$

is selected such that $\sqrt{\lambda} = \omega > 0$. This represents an outgoing wave if we adjoin the factor $e^{-i\omega t}$.

(iii) When λ is real and negative, we again choose $\sqrt{\lambda}$ in (3) to be the square root of λ that has a positive imaginary part for the same reason as in (i). For the particular case $\lambda = -k^2$, where k is real and positive, the formula (3) becomes

$$E(\mathbf{x};\boldsymbol{\xi}) = e^{-kr}/4\pi r . \tag{5}$$

The solutions which correspond to (3), (4), and (5) in two dimensions are

$$(i/4) H_0^{(1)}(|\mathbf{x} - \xi| \sqrt{\lambda}), \qquad (i/4) H_0^{(1)}(\omega |\mathbf{x} - \xi|), \qquad (1/2\pi) K_0(k |\mathbf{x} - \xi|),$$

respectively. Here, the functions $H_0^{(1)}$ and K_0 are the Hankel and modified Bessel functions, respectively.

The integral representation formula for the solution of the inhomogeneous equation

$$(\nabla^2 - k^2) u = -4\pi \rho \tag{6}$$

is obtained from the relations (2) and (6) by using Green's identity and is readily found to be

$$u(\mathbf{x}) = \int_R \frac{\rho e^{-kr}}{r} dV + \frac{1}{4\pi} \int_S u \frac{\partial}{\partial n} \left(\frac{e^{-kr}}{r} \right) dS - \frac{1}{4\pi} \int_S \frac{e^{-kr}}{r} \frac{\partial u}{\partial n} dS . \tag{7}$$

The interpretation of these integrals as volume, single-layer, and double-layer potentials is the same as for the corresponding formulas in Section 6.2. The properties of these potentials are also similar. For instance, the formulas that correspond to (6.2.7) and (6.2.8) are

$$\frac{\partial u}{\partial n} \bigg|_{P_\pm} = \mp 2\pi\sigma(P) + \int_S \sigma(Q) \frac{\partial}{\partial n} \left(\frac{e^{-kr}}{r} \right) dS , \tag{8}$$

where

$$u = \int_S \sigma(Q) \left(\frac{e^{-kr}}{r} \right) dS . \tag{9}$$

Similarly, the formulas that correspond to (6.2.10) and (6.2.11) are

$$u|_{P_\pm} = \pm 2\pi\tau(P) + \int_S \tau(Q) \frac{\partial}{\partial n} (e^{-kr}/r) \, dS , \tag{10}$$

where

$$u = \int_S \tau(Q) \frac{\partial}{\partial n} (e^{-kr}/r) \, dS . \tag{11}$$

The rest of the notation is the same as in Section 6.2.

The integral representation of solutions of the exterior and interior Dirichlet and Neumann problems is achieved in an analogous manner, as shall become evident from the examples of physical interest which are presented in the next section. We end this section by mentioning the so-called Sommerfeld radiation condition. A three-dimensional solution of the Helmholtz equation $(\nabla^2 + k^2)u = 0$ is said to satisfy the radiation condition if

$$\lim r\left(\frac{\partial u}{\partial r} - iku\right) = 0 , \tag{12}$$

as $r \to \infty$. Physically, this condition implies that there are no incoming waves from infinity. In two dimensions, the corresponding condition is

$$\lim \sqrt{r}\left(\frac{\partial u}{\partial r} - iku\right) = 0 , \tag{13}$$

as $r \to \infty$. The free-space Green's function satisfies the radiation condition.

6.7. EXAMPLES

Example 1. *Acoustic diffraction of a plane wave by a perfectly soft disk.* We follow the coordinate system and notation of Example 1 in Section 6.3. Furthermore, we assume the time dependence of the form $e^{-i\omega t}$ for the wave functions involved in the problem and omit this factor in the sequel. The time-independent part of the velocity potential u is

$$u(\rho, \varphi, z) = u_i(\rho, \varphi, z) + u_s(\rho, \varphi, z) , \tag{1}$$

where u_i and u_s denote the velocity potentials of the incident and diffracted fields. All three functions occurring in equation (1) satisfy the Helmholtz equation. The boundary value problem is

$$(\nabla^2 + k^2)u_s = 0 , \tag{2}$$

$$u_i(\rho \ \varphi, 0) + u_s(\rho, \varphi, 0) = 0 , \qquad 0 \leqslant \rho \leqslant a , \tag{3}$$

u_s and $\partial u_s/\partial z$ are continuous across $z = 0$, $a < \rho < \infty$, (4)

and u_s satisfies the radiation condition at infinity. The fundamental solution E that satisfies the equation

$$-(\nabla^2 + k^2)E = \delta(\mathbf{x} - \boldsymbol{\xi}) \tag{5}$$

as well as the radiation condition is

$$E(\mathbf{x};\boldsymbol{\xi}) = \frac{\exp(ik|\mathbf{x}-\boldsymbol{\xi}|)}{4\pi|\mathbf{x}-\boldsymbol{\xi}|}$$

$$= \frac{\exp ik\{\rho^2 + t^2 - 2\rho t[\cos(\varphi-\varphi_1)] + (z-z_1)^2\}^{\frac{1}{2}}}{4\pi\{\rho^2 + t^2 - 2\rho t[\cos(\varphi-\varphi_1)] + (z-z_1)^2\}^{\frac{1}{2}}}. \tag{6}$$

Multiply (2) by E, (5) by u_s, subtract, integrate, use Green's second identity, and obtain

$$u_s(\mathbf{x}) = \int_{S^+ + S^-} \left(E\frac{\partial u_s}{\partial n} - u_s\frac{\partial E}{\partial n} \right) dS, \tag{7}$$

where S^+ and S^- are the upper and lower surfaces of the disk, respectively. On $S^\pm$, we have $\partial/\partial n = \mp\partial/\partial z$, respectively. Thus, (7) can be written as

$$u_s(\rho,\varphi,z) = \int_0^a\int_0^{2\pi}\left[-E\frac{\partial u_s}{\partial z_1} + u_s\frac{\partial E}{\partial z_1} \right]_{z_1=0+} t\,d\varphi_1\,dt$$

$$+ \int_0^a\int_0^{2\pi}\left[E\frac{\partial u_s}{\partial z_1} - u_s\frac{\partial E}{\partial z_1} \right]_{z_1=0-} t\,d\varphi_1\,dt$$

$$= -\int_0^a\int_0^{2\pi} t\sigma(t,\varphi_1,0)\,E|_{z_1=0}\,d\varphi_1\,dt, \tag{8}$$

where we have used the fact that $u_s = -u_i$ on both sides of the disk and where

$$\sigma(t,\varphi_1,0) = \left(\frac{\partial u_s}{\partial z}\bigg|_{z_1=0+} - \frac{\partial u_s}{\partial z}\bigg|_{z_1=0-} \right). \tag{9}$$

When we apply the boundary condition (3) in (8), we get

$$u_i(\rho,\varphi,0) = \frac{1}{4\pi}\int_0^a\int_0^{2\pi} \frac{t\sigma(t,\varphi_1,0)\exp ik[\rho^2 + t^2 - 2\rho t\cos(\varphi-\varphi_1)]^{\frac{1}{2}}}{[\rho^2 + t^2 - 2\rho t\cos(\varphi-\varphi_1)]^{\frac{1}{2}}}$$

$$\times d\varphi_1\,dt. \tag{10}$$

In view of the Fourier superposition principle, we can assume that $u_i(\rho, \varphi, z)$ and $\sigma(\rho, \varphi)$ are of the form $u_i^{(n)}(\rho, z) \cos n\varphi$ and $2\sigma^{(n)}(\rho) \cos n\varphi$, respectively. Then, proceeding as in the steps that led from (6.3.2) to (6.3.4), we obtain from (10) the integral equation

$$u_i^{(n)}(\rho, 0) = \int_0^a t\sigma^{(n)}(t) K_1(t, \rho) \, dt , \qquad (11)$$

where the kernel $K_1(t, \rho)$ is

$$K_1(t, \rho) = \frac{1}{2\pi} \int_0^{2\pi} \frac{\exp ik(\rho^2 + t^2 - 2\rho \cos \psi)^{\frac{1}{2}}}{(\rho^2 + t^2 - 2\rho t \cos \psi)^{\frac{1}{2}}} \cos n\psi \, d\psi . \qquad (12)$$

The integral equation (11) is the required Fredholm integral equation of the first kind which embodies the solution of the boundary value problem (2)–(4).

For an annular disk with inner radius b and outer radius a, the corresponding integral equation is

$$u_i^{(n)}(\rho, 0) = \int_b^a t\sigma^{(n)}(t) K_1(t, \rho) \, dt . \qquad (13)$$

Example 2. *Torsional oscillations of an elastic half-space.* In terms of cylindrical polar coordinates, the axially symmetric boundary value problem

$$\frac{\partial^2 v}{\partial \rho^2} + \frac{1}{\rho} \frac{\partial v}{\partial \rho} - \frac{v}{\rho^2} + \frac{\partial^2 v}{\partial z^2} + k^2 v = 0 , \qquad k^2 = \frac{\omega^2 da^2}{\mu} , \qquad (14)$$

$$v = \Omega\rho , \qquad z = 0 , \qquad 0 \leqslant \rho \leqslant 1 , \qquad (15)$$

$$\partial v/\partial z = 0 , \qquad z = 0 , \qquad \rho > 1 , \qquad (16)$$

embodies the torsional oscillations of a homogeneous and isotropic half-space which occupies the region $z \geqslant 0$. A disk of radius a is attached to it and is forced to execute torsional oscillations with period $2\pi/\omega$. All lengths are made dimensionless with a as the standard length, so that the disk occupies the region $z = 0$, $0 \leqslant \rho \leqslant 1$. The quantities d and μ are respectively the density and shear modulus of the elastic material, while k is a dimensionless parameter.

It is easily verified from (14) that the function $w(\rho, \varphi, z)$,

$$w(\rho, \varphi, z) = v(\rho, z)\cos\varphi , \tag{17}$$

satisfies the Helmholtz equation

$$(\nabla^2 + k^2)w = 0 . \tag{18}$$

The Green's function that corresponds to this boundary value problem satisfies the auxiliary system

$$(\nabla^2 + k^2)G(\mathbf{x}; \boldsymbol{\xi}) = -4\pi\delta(\mathbf{x} - \boldsymbol{\xi}) ; \quad \frac{\partial G}{\partial z_1}\bigg|_{z_1 = 0} = 0 , \tag{19}$$

where, as before, $\mathbf{x} = (\rho, \varphi, z)$ and $\boldsymbol{\xi} = (t, \varphi_1, z_1)$. Again the method of images gives us the value of G

$$G(\mathbf{x}; \boldsymbol{\xi}) = \frac{\exp ikr}{r} + \frac{\exp ikr'}{r'} , \tag{20}$$

where $r = |\mathbf{x} - \boldsymbol{\xi}|$, $r' = |\mathbf{x} - \boldsymbol{\xi}'|$, and $\boldsymbol{\xi}'$ is the image of $\boldsymbol{\xi}$ in the plane $z = 0$, that is, $\boldsymbol{\xi}' = (t, \varphi_1, -z_1)$. Thus,

$$\begin{aligned}
r &= \{\rho^2 + t^2 - 2\rho t[\cos(\varphi - \varphi_1)] + (z - z_1)^2\}^{1/2} , \\
r' &= \{\rho^2 + t^2 - 2\rho t[\cos(\varphi - \varphi_1)] + (z - z_1)^2\}^{1/2} .
\end{aligned} \tag{21}$$

The differential equations (14) and (19) present us with the integral representation formula by the routine procedure. The required relation is

$$\begin{aligned}
w &= v(\rho, z)\cos\varphi \\
&= \frac{1}{4\pi} \int\limits_{z_1 = 0} \left[-G\frac{\partial v}{\partial z_1}\cos\varphi_1 + v\frac{\partial G}{\partial z_1}\cos\varphi_1 \right]_{z_1 = 0} dS .
\end{aligned} \tag{22}$$

Applying the boundary conditions (15), (16), and $(19)_2$ in (22) and using (20), we have

$$\Omega\rho\cos\varphi = \frac{1}{2\pi} \int\limits_0^1 t\phi(t)$$

$$\times \int\limits_0^{2\pi} \frac{\exp ik[\rho^2 + t^2 - 2\rho t\cos(\varphi - \varphi_1)]^{1/2}}{[\rho^2 + t^2 - 2\rho t\cos(\varphi - \varphi_1)]^{1/2}} \cos\varphi_1 \, d\varphi_1 \, dt, \tag{23}$$

where

$$\phi(t) = -\frac{\partial v}{\partial z_1}\bigg|_{z_1=0} . \tag{24}$$

Set $\varphi_1 - \varphi = \psi$, proceed as we did in the relations (6.3.2)–(6.3.4), and obtain

$$\Omega\rho = \int_0^1 t\phi(t) K_1(t,\rho)\, dt , \qquad 0 < \rho < 1 , \tag{25}$$

where

$$K_1(t,\rho) = \frac{1}{2\pi} \int_0^{2\pi} \frac{\exp ik(\rho^2+t^2-2\rho t\cos\psi)^{\frac12}}{(\rho^2+t^2-2\rho t\cos\psi)^{\frac12}} \cos\psi\, d\psi . \tag{26}$$

Finally, we use the identities [21]

$$\frac{\exp ik(\rho^2+t^2-2\rho t\cos\psi)^{\frac12}}{(\rho^2+t^2-2\rho\cos\psi)^{\frac12}} = \int_0^\infty \frac{pJ_0[p(\rho^2+t^2-2\rho t\cos\psi)^{\frac12}]}{\gamma}\, dp, \tag{27}$$

where

$$\gamma = \begin{cases} -i(k^2-p^2)^{\frac12} , & k \geqslant p \\ (p^2-k^2)^{\frac12} , & p \geqslant k \end{cases}$$

and

$$J_0[p(\rho^2+t^2-2\rho t\cos\psi)^{\frac12}] = \sum_{r=0}^\infty (2-\delta_{0r})(\cos r\psi) J_r(p\rho) J_r(pt) \tag{28}$$

in (26), use the orthogonality of cosine functions, and derive

$$K_1(t,\rho) = \int_0^\infty (p/\gamma) J_1(p\rho) J_1(pt)\, dp . \tag{29}$$

When $\omega \to 0$, we get the corresponding integral equation for the steady rotation of the elastic half-space, and (29) becomes

$$K_1(t,\rho) = \int_0^\infty J_1(p\rho) J_1(pt)\, dp . \tag{30}$$

For the case of an annular disk with inner radius b and outer radius a, the integral equation that corresponds to (25) is of the form

$$\Omega\rho = \int_b^a t\phi(t) K_1(t,\rho) dt , \qquad b < \rho < a . \tag{31}$$

Example 3. *Steady Stokes flow in an unbounded medium.* The Stokes flow equations

$$\nabla^2 \mathbf{q} = \nabla p , \qquad \nabla \cdot \mathbf{q} = 0 \tag{32}$$

govern the slow, steady flow of incompressible viscous fluids. These equations have been made dimensionless with the help of the free-stream velocity u and a characteristic length a inherent in the problem. The quantities $\mathbf{q}$ and p stand for the velocity vector and pressure, respectively.

Let S be the surface of a solid B moving in the fluid; then, the boundary conditions are

$$\mathbf{q}(x) = \mathbf{e}_1 , \qquad \mathbf{x} \in S , \qquad \mathbf{q}(\mathbf{x}) = 0 \quad \text{as} \quad \mathbf{x} \to \infty , \tag{33}$$

where $\mathbf{e}_1$ is the direction of motion of B, taken to be in the x_1 direction. The boundary value problem (32)–(33) can be converted into a Fredholm integral equation of the first kind by defining the Green's tensor $\mathbf{T}_1$ (or T_{1ik}) and Green's vector $\mathbf{p}_1$ (or p_{1j}), which satisfy the mathematical system

$$\nabla^2 \mathbf{T}_1 - \nabla \mathbf{p}_1 = -\mathbf{I}\delta(\mathbf{x}-\xi) , \tag{34}$$

$$\nabla \cdot \mathbf{T}_1 = 0 , \qquad \mathbf{T}_1 \to 0 \quad \text{as} \quad \mathbf{x} \to \infty , \tag{35}$$

where $\mathbf{I} = \delta_{ij}$, the Kronecker delta.

It follows by direct verification that the system (34)–(35) has the representation formulas

$$\mathbf{T}_1 = (1/8\pi)[\mathbf{I}\nabla^2 \phi - \text{grad grad } \phi] , \qquad \mathbf{p}_1 = -(1/8\pi)\text{grad } \nabla^2 \phi , \tag{36}$$

$$\nabla^4 \phi = \nabla^2 \nabla^2 \phi = -8\pi \delta(\mathbf{x}-\xi) . \tag{37}$$

The appropriate solution of the biharmonic equation (37) is $\phi = r = |\mathbf{x}-\xi|$. Thus,

$$\mathbf{T}_1 = (1/8\pi)[\mathbf{I}\nabla^2 |\mathbf{x}-\xi| - \text{grad grad } |\mathbf{x}-\xi|] , \tag{38}$$

$$\mathbf{p}_1 = -(1/8\pi)\text{grad } \nabla^2 |\mathbf{x}-\xi| . \tag{39}$$

The required integral equation formula now follows by taking the scalar product of (32) by $\mathbf{T}_1$ and of (34) by $\mathbf{q}$ and using the usual steps of subtracting and integrating. In the integral so obtained, there occur terms $\mathbf{T} \cdot \nabla p$ and $\mathbf{q} \cdot \nabla \mathbf{p}$ and they can be processed by using the identities

$$\nabla \cdot (\mathbf{q}\mathbf{p}_1) = \mathbf{q} \cdot \nabla \mathbf{p}_1 \,, \qquad \nabla \cdot (p\mathbf{T}_1) = \mathbf{T}_1 \cdot \nabla p \,, \tag{40}$$

where we have used the results $\nabla \cdot \mathbf{q} = \nabla \cdot \mathbf{T}_1 = 0$. The final result is

$$\mathbf{q}(\mathbf{x}) = -\int_S \left[\left(\frac{\partial \mathbf{q}}{\partial n} - p\mathbf{n} \right) \cdot \mathbf{T}_1 - \mathbf{q} \cdot \left(\frac{\partial \mathbf{T}_1}{\partial n} - \mathbf{n}\mathbf{p}_1 \right) \right] dS \,. \tag{41}$$

From equation (34) and the divergence theorem, it follows that, if $\mathbf{q}$ is constant on S, then equation (41) reduces to

$$\mathbf{q}(\mathbf{x}) = -\int_S \mathbf{f} \cdot \mathbf{T}_1 \, dS \,, \tag{42}$$

where

$$\mathbf{f} = \frac{\partial \mathbf{q}}{\partial n} - p\mathbf{n} \,. \tag{43}$$

Finally, using the boundary condition $(33)_1$, we have the required integral equation

$$\mathbf{e}_1 = -\int_S \mathbf{f} \cdot \mathbf{T}_1 \, dS \,. \tag{44}$$

Example 4. *Steady Oseen flow.* The dimensionless Oseen equations are

$$\mathcal{R} \, \partial \mathbf{q} / \partial x_1 = -\nabla p + \nabla^2 \mathbf{q} \,, \qquad \nabla \cdot \mathbf{q} = 0 \,, \tag{45}$$

$$\mathbf{q} = \mathbf{e}_1 \quad \text{on} \quad S \,; \quad \mathbf{q} \to 0 \quad \text{as} \quad \mathbf{x} \to \infty \,, \tag{46}$$

where $\mathcal{R} = ua/v$ is the Reynolds number, v is the coefficient of kinematic viscosity, and a and u are the same quantities as defined in the previous example. The Green's tensor $\mathbf{T}$ and Green's vector $\mathbf{p}$ for this problem satisfy the system

$$\mathcal{R} \, \partial \mathbf{T} / \partial x_1 = -\nabla \mathbf{p} + \nabla^2 \mathbf{T} + \mathbf{I}\delta(\mathbf{x} - \xi) \,, \tag{47}$$

$$\nabla \cdot \mathbf{T} = 0 \,, \qquad \mathbf{T} \to 0 \quad \text{as} \quad \mathbf{x} \to \infty \,. \tag{48}$$

The corresponding representation formulas are

$$\mathbf{T} = (1/8\pi) [\mathbf{I} \nabla^2 \phi - \operatorname{grad} \operatorname{grad} \phi] \,, \tag{49}$$

$$\mathbf{p} = -(1/8\pi)\,\text{grad}\,[\nabla^2\phi - \mathscr{R}\,\partial\phi/\partial x_1]\,. \tag{50}$$

$$\nabla^2(\nabla^2 - \mathscr{R}\,\partial/\partial x_1)\phi = -8\pi\,\delta(\mathbf{x}-\xi)\,. \tag{51}$$

To solve equation (51) for ϕ, we first use the formula

$$\nabla^2(1/|\mathbf{x}-\xi|) = -4\pi\,\delta(\mathbf{x}-\xi) \tag{52}$$

in it and obtain

$$\nabla^2(\nabla^2 - \mathscr{R}\,\partial/\partial x_1)\phi = 2\nabla^2(1/|\mathbf{x}-\xi|)\,. \tag{53}$$

Thus, if ϕ satisfies

$$\nabla^2\phi - \mathscr{R}\,\partial\phi/\partial x_1 = 2/|\mathbf{x}-\xi|\,, \tag{54}$$

then (51) is satisfied.

On the other hand, we can also write (51) as

$$(\nabla^2 - \mathscr{R}\,\partial/\partial x_1)\nabla^2\phi = -8\pi\,\delta(\mathbf{x}-\xi)\,. \tag{55}$$

By setting

$$\nabla^2\phi = \psi \tag{56}$$

and using the identity

$$(\nabla^2 - \sigma^2)[e^{-\sigma(x_1-\xi_1)}\psi] = e^{-\sigma(x_1-\xi_1)}[\nabla^2 - 2\sigma\,\partial/\partial x_1]\psi\,, \tag{57}$$

we can write (55) as

$$(\nabla^2 - \sigma^2)[e^{-\sigma(x_1-\xi_1)}\psi] = -8\pi e^{-\sigma(x_1-\xi_1)}\delta(\mathbf{x}-\xi) \tag{58}$$

with $\sigma = 2\mathscr{R}$. Now, observe that, by the nature of the Dirac delta function, the factor $e^{-\sigma(x_1-\xi_1)}$ influences the equation only at $x_1 = \xi_1$, where its value is unity. Thus, (58) yields

$$\psi(\mathbf{x},\xi) = \nabla^2\phi = \frac{2}{|\mathbf{x}-\xi|}\exp\left\{-|\sigma|\left[|\mathbf{x}-\xi| - \frac{\sigma}{|\sigma|}(x_1-\xi_1)\right]\right\}. \tag{59}$$

From (54) and (59), we obtain

$$\frac{\partial\phi}{\partial x_1} = -\frac{1}{\sigma|\mathbf{x}-\xi|}\left(1 - \exp\left\{-|\sigma|\left[|\mathbf{x}-\xi| - \frac{\sigma}{|\sigma|}(x_1-\xi_1)\right]\right\}\right). \tag{60}$$

If we set

$$s = |\mathbf{x}-\xi| - (\sigma/|\sigma|)(x_1-\xi_1)\,, \tag{61}$$

we have

$$\frac{\partial s}{\partial x_1} = \frac{(x_1 - \xi_1)}{|\mathbf{x} - \boldsymbol{\xi}|} - \frac{\sigma}{|\sigma|} = \frac{(x_1 - \xi_1)}{|\mathbf{x} - \boldsymbol{\xi}|} - \frac{|\sigma|}{\sigma},$$

$$= -\frac{|\sigma|}{\sigma|\mathbf{x} - \boldsymbol{\xi}|}\left(|\mathbf{x} - \boldsymbol{\xi}| - \frac{\sigma}{|\sigma|}(x_1 - \xi_1)\right) = -\frac{|\sigma| s}{\sigma|\mathbf{x} - \boldsymbol{\xi}|}. \quad (62)$$

Combining (60)–(62) with

$$\frac{\partial \phi}{\partial x_1} = \frac{\partial \phi}{\partial s}\frac{\partial s}{\partial x_1},$$

we find that

$$d\phi/ds = (1 - e^{-|\sigma|s})/|\sigma| s$$

or

$$\phi = (1/|\sigma|) \int_0^{|\sigma|s} [(1 - e^{-t})/t]\, dt. \quad (63)$$

Thereby, the Green's tensor $\mathbf{T}$ and Green's vector $\mathbf{p}$ are determined.

The integral equation equivalent to the boundary value problem (45)–(46) is now found precisely the way we found the integral equation (41). Indeed, the present formula is

$$\mathbf{q}(\mathbf{x}) = -\int_S \left[\mathbf{T}\cdot\left(\frac{\partial \mathbf{q}}{\partial n} - p\mathbf{n}\right) - \left(\frac{\partial \mathbf{T}}{\partial n} - p\mathbf{n}\right)\cdot\mathbf{q} - \mathscr{R}(\mathbf{T}\cdot\mathbf{q})n_1\right] dS, \quad (64)$$

where n_1 is the x_1 component of the outward normal. When we use the boundary condition $\mathbf{q} = \mathbf{e}_1$ on S, equation (47), and the divergence theorem, we obtain

$$\mathbf{e}_1 = -\int_S \mathbf{T}\cdot\mathbf{f}\, dS, \quad (65)$$

where $\mathbf{f}$ is defined by (43).

Example 5. *Heat conduction.* The boundary value problem

$$\nabla^2 u - k^2 u = -\rho(\mathbf{x}), \qquad \mathbf{x} \in R_e, \qquad \left.\frac{\partial u}{\partial n}\right|_S = 0 \quad (66)$$

embodies the solution of the heat conduction problem of an infinite expanse of material containing a cavity S on which the temperature gradient is zero. Our aim is to give the system (66) an integral-equation formulation. For this purpose, we assume that u can be represented as the sum of a volume potential and a single-layer potential

$$4\pi u(\mathbf{x}) = \int_{R_e} \rho(\xi)\, E(\mathbf{x};\xi)\, dV + \int_S \sigma(\xi)\, E(\mathbf{x};\xi)\, dS\,, \tag{67}$$

where $E(\mathbf{x};\xi) = [\exp(-k|\mathbf{x}-\xi|)]/|\mathbf{x}-\xi|$ and σ is an unknown source density.

The next step is to take the normal derivative of both sides of equation (67), let $\mathbf{x}$ approach S, and use relation (6.6.8). The result is

$$0 = \int_{R_e} \rho(\xi)\frac{\partial E(\mathbf{x};\xi)}{\partial n}\, dV - 2\pi\sigma(\mathbf{x}) + \int_S \sigma(\xi)\frac{\partial E(\mathbf{x};\xi)}{\partial n}\, dS\,, \tag{68}$$

which is a Fredholm integral equation of the second kind in $\sigma(\mathbf{x})$.

The reader who is familiar with the theory of heat conduction is advised to formulate the corresponding problem of the composite medium into a Fredholm integral equation of the first kind.

EXERCISES

1. Show that when S is the surface of a unit sphere (a) the solution of the integral equation

$$\rho z \cos\varphi = \int_S \sigma(\rho_1, z_1)\, |\mathbf{x}-\xi|^{-1}\cos\varphi_1\, dS$$

is $\sigma = (5/12)P_2^{\,1}(\cos\theta)$; (b) the solution of the integral equation

$$(1/2)\rho z^2 \cos\varphi = (3/8\pi)\int_S P_1^{\,1}(\cos\theta_1)\, |\mathbf{x}-\xi|\cos\varphi_1\, dS$$

$$+ \int_S \sigma(\rho_1, z_1)\, |\mathbf{x}-\xi|^{-1}\cos\varphi_1\, dS$$

is $\sigma = (1/4\pi)[(3/2)P_1^{\,1}(\cos\theta) + (7/15)P_3^{\,1}(\cos\theta)]$.

In the above relations, (ρ, φ, z) are cylindrical polar coordinates.

Hints: (i) Use the formula

$$|\mathbf{x}-\boldsymbol{\xi}| = r_> \sum_{v=1}^{\infty} \left(\frac{x^2}{2v+3} - \frac{1}{2v-1} \right) x^v P_v(\cos \gamma) ,$$

where

$$x = r_</r_> , \qquad r_< = \min(r, r_1) , \qquad r_> = \max(r, r_1) ,$$

$$\cos \gamma = \cos \theta \cos \theta_1 + \sin \theta \sin \theta_1 \cos(\varphi - \varphi_1) ,$$

while (r, φ, θ) and $(r_1, \varphi_1, \theta_1)$ are the spherical polar coordinates of $\mathbf{x}$ and $\boldsymbol{\xi}$, respectively.

(ii) On the surface of the unit sphere, $\rho = \sin \theta$, $z = \cos \theta$.

2. Show that the solutions of the integral equations

$$\rho \cos \varphi = \int_S \sigma(\rho_1, z_1) |\mathbf{x}-\boldsymbol{\xi}|^{-1} \cos \varphi_1 \, dS$$

and

$$0 = \int_S \sigma'(\rho_1, z_1) |\mathbf{x}-\boldsymbol{\xi}|^{-1} \cos \varphi_1 \, dS$$

$$+ \tfrac{1}{2} \int_S \sigma(\rho_1, z_1) |\mathbf{x}-\boldsymbol{\xi}| \cos \varphi_1 \, dS ,$$

where S is the surface of a thin circular disk of unit radius, are

$$\sigma = 2\rho/\pi^2 (1-\rho^2)^{\frac{1}{2}} , \qquad \sigma' = \rho(2-\rho^2)/3\pi^2 (1-\rho^2)^{\frac{1}{2}} .$$

3. Starting with the Cauchy integral formula for an analytic function

$$f(z) = (1/2\pi i) \int_C [f(t)/(t-z)] \, dt ,$$

where C is the circumference $|z| = a$ and z is in the interior of C, and using the formula

$$0 = (1/2\pi i) \int_C [f(t)/(t-z^*)] \, dt , \qquad z^* = a^2/\bar{z} ,$$

which is a result of Cauchy theorem because the image part z^* (of z) is exterior to C, derive the Poisson integral formula in a plane:

$$u(\rho, \varphi) = \frac{a^2 - \rho^2}{2\pi} \int_0^{2\pi} \frac{f(a, \theta) \, d\theta}{a^2 - 2a\rho [\cos(\theta - \varphi)] + a^2} .$$

4. Discuss the single-layer and double-layer potentials in two-dimensional potential theory by starting with the formula

$$E(\mathbf{x};\xi) = (1/2\pi)\log(1/|\mathbf{x}-\xi|)$$

instead of (6.2.2).

5. With the help of the results obtained in the previous exercise, prove that the solution of the interior Dirichlet problem in two dimensions can be written as

$$w(\mathbf{x}) = \int_C [(\cos\psi)/|\mathbf{x}-\xi|]\,\tau(\xi)\,d\ell\;,$$

where ψ is the angle between $(\mathbf{x}-\xi)$ and $\mathbf{n}$; and $d\ell$ denotes the element of the arc length along the curve C.

6. Proceeding as in Example 1 of Section 6.7, show that the following boundary value problem,

$$(\nabla^2+k^2)u_s = 0\;,$$

$$\partial u_s/\partial z = -\partial u_i/\partial z \quad \text{on} \quad z=0\;, \quad 0\leqslant\rho\leqslant a\;,$$

u_s and $\partial u_s/\partial z$ are continuous across $z=0$, $\quad a<\rho<\infty$,

and u_s satisfies the radiation condition, reduces to the integral equation

$$\frac{\partial u_i}{\partial z}(\rho,\varphi,0) = \int_0^a t\sigma(t)\left[\frac{\partial^2}{\partial z^2}\left(\frac{\exp ik\,|\mathbf{x}-\xi|}{|\mathbf{x}-\xi|}\right)\right]_{z_1=z=0} d\varphi_1\,dt\;,$$

where $\sigma(t)=(1/4\pi)\big(u_s|_{z_1=0+}-u_s|_{z=0-}\big)$. This boundary value problem embodies the solution of the diffraction of a plane wave by a perfectly rigid, circular thin disk.

7. By following the method and notation of Example 3 in Section 6.7, prove that the integral representation formula for the velocity vector when the fluid is bounded by a vessel Σ is

$$\mathbf{q}(\mathbf{x}) = -\int\left[\left(\frac{\partial\mathbf{q}}{\partial n}-p\mathbf{n}\right)\cdot\mathbf{T}-\mathbf{q}\cdot\left(\frac{\partial\mathbf{T}}{\partial n}-\mathbf{n}p\right)\right]dS\;.$$

Substituting the boundary condition $\mathbf{q}=\mathbf{e}_1$ on S gives the Fredholm integral equation

$$\mathbf{e}_1 = -\int_S \mathbf{f} \cdot \mathbf{T}\, dS .$$

Here, $\mathbf{T}$ satisfies the boundary value problem

$$\nabla^2 \mathbf{T} - \operatorname{grad} p = \mathbf{I}\delta(\mathbf{x}-\boldsymbol{\xi}) , \qquad \nabla\cdot\mathbf{T} = 0 , \qquad \mathbf{T} = 0 \quad \text{on} \quad \Sigma .$$

8. The dimensionless equations of elastostatics are

$$(\lambda+\mu)\operatorname{grad}\theta + \mu\nabla^2\mathbf{u} = 0 , \qquad \theta = \operatorname{div}\mathbf{u} ,$$

where $\mathbf{u}$ (u_i. $i = 1, 2, 3$) is the displacement vector, and λ and μ are Lamé constants of the medium. The above equations have been made dimensionless by introducing a characteristic geometric length a.

Consider the uniform translation of a light, rigid body B with surface S_1, which is embedded in an infinite homogeneous and isotropic medium. Prove that the corresponding Green's tensor $\mathbf{T}_1$ and Green's dilation vector θ_1 are given by the formulas

$$T_{1ij} = \frac{1}{8\pi}\left[\frac{\lambda+3\mu}{\lambda+2\mu}\frac{\delta_{ij}}{|\mathbf{x}-\boldsymbol{\xi}|} + \frac{\lambda+\mu}{\lambda+2\mu}\frac{(x_i-\xi_i)(x_j-\xi_j)}{|\mathbf{x}-\boldsymbol{\xi}|^3}\right] ,$$

$$\theta_{1i} = -\frac{1}{4\pi}\frac{\mu}{\lambda+2\mu}\left[\frac{x_i-\xi_i}{|\mathbf{x}-\boldsymbol{\xi}|^3}\right] .$$

Also prove that the integral representation formula is

$$\mathbf{u}(p) = -\int_S \left\{\left[\mu\frac{d\mathbf{u}}{\partial n} + (\lambda+\mu)\theta\mathbf{n}\right]\cdot\mathbf{T}\right.$$

$$\left. -\mathbf{u}\cdot\left[\mu\frac{d\mathbf{T}_1}{dn} + (\lambda+\mu)\theta_1\,\mathbf{n}\right]\right\}dS .$$

Using the boundary conditions

$$\mathbf{u} = (d_0/a)\mathbf{e} \qquad \text{on} \quad S ; \qquad \mathbf{u} \to 0 \quad \text{at} \quad \infty ;$$

where d_0 is the magnitude of the translation and $\mathbf{e}$ is the unit direction along the translation of B, show that the Fredholm integral equation that embodies the solution of the above problem is

$$(d_0/a)\mathbf{e} = -\int \mathbf{f}\cdot\mathbf{T}_1\, dS ,$$

where

$$\mathbf{f} = \mu(d\mathbf{u}/dn) + (\lambda+\mu)\theta\mathbf{n} .$$

Extend the above results to a bounded medium.

9. Proceed as in Example 1 of Section 6.3 and Example 1 of Section 6.7 and give the integral-equation formulation of these problems for a spherical cap.

10. The Schrödinger equation

$$\nabla^2 \psi(\mathbf{x}) - (2m/\hbar^2)V(\mathbf{x})\psi(\mathbf{x}) + k^2\psi(\mathbf{x}) = 0$$

with boundary condition that $\psi(\mathbf{x})\exp(-iEt/\hbar)$ represents an incident plane wave, with wave vector $\mathbf{k}_0$, as well as the condition that we have an outgoing wave as $\mathbf{x} \to \infty$, describes the quantum mechanical theory of scattering by a potential $V(\mathbf{x})$. Here, $k^2 = k_0{}^2 = 2mE/\hbar^2$, and other quantities have their usual meaning.

Use the method of Section 6.6 and prove that we can transform this scattering problem into the integral equation

$$\psi(\mathbf{x}) = (\exp i\mathbf{k}_0 \cdot \mathbf{x}) - \frac{m}{2\pi\hbar^2} \int_R \frac{\exp ik|\mathbf{x}-\boldsymbol{\xi}|}{|\mathbf{x}-\boldsymbol{\xi}|} V(\boldsymbol{\xi})\psi(\boldsymbol{\xi})\, dV.$$

Find its solution by the method of successive approximations (the approximation obtained in this way is called the Born approximation).

SYMMETRIC KERNELS

7.1. INTRODUCTION

A kernel $K(s, t)$ is symmetric (or complex symmetric or Hermitian) if

$$K(s, t) = K^*(t, s) , \tag{1}$$

where the asterisk denotes the complex conjugate. In the case of a real kernel, the symmetry reduces to the equality

$$K(s, t) = K(t, s) . \tag{2}$$

We have seen in the previous two chapters that the integral equations with symmetric kernels are of frequent occurrence in the formulation of physically motivated problems.

We claim that if a kernel is symmetric then all its iterated kernels are also symmetric. Indeed,

$$K_2(s, t) = \int K(s, x) K(x, t) \, dx = \int K^*(t, x) K^*(x, s) \, dx = K_2^*(t, s) .$$

Again, if $K_n(s, t)$ is symmetric, then the recursion relation gives

$$K_{n+1}(s, t) = \int K(s, x) K_n(x, t) \, dx$$

$$= \int K_n^*(t, x) K^*(x, s) \, dx = K_{n+1}^*(t, s) . \tag{3}$$

The proof of our claim follows by induction. Note that the *trace* $K(s,s)$ of a symmetric kernel is always real because $K(s,s) = K^*(s,s)$. Similarly, the traces of all iterates are also real.

Complex Hilbert Space

We present here a brief review of the properties of the complex Hilbert space $\mathscr{L}_2(a,b)$, which shall be needed in the sequel. This discussion is valid for real $\mathscr{L}_2$-space as a special case. A linear space of infinite dimension with inner product (or scalar product) (x,y) which is a complex number satisfying (a) the definiteness axiom $(x,x) > 0$ for $x \neq 0$; (b) the linearity axiom $(\alpha x_1 + \beta x_2, y) = \alpha(x_1,y) + \beta(x_2,y)$, where α and β are arbitrary complex numbers; and (c) the axiom of (Hermitian) symmetry $(y,x) = (x,y)^*$; is called a complex Hilbert space.

Let H be the set of complex-valued functions $\phi(t)$ defined in the interval (a,b) such that

$$\int |\phi(t)|^2 \, dt < \infty . \tag{4}$$

Furthermore, let us define the inner product by the bilinear form (1.6.1):

$$(\phi, \psi) = \int \phi(t)\psi^*(t)\, dt . \tag{5}$$

Then, H is a linear and complex Hilbert space $\mathscr{L}_2(a,b)$ (or $\mathscr{L}_2$). The norm $\|\phi\|$ as defined by (1.6.2),

$$\|\phi\| = \left(\int |\phi(t)|^2 \, dt \right)^{\frac{1}{2}} \tag{6}$$

is called the norm that generates the natural metric

$$d(\phi, \psi) = \|\phi - \psi\| = (\phi - \psi, \phi - \psi)^{\frac{1}{2}} . \tag{7}$$

The Schwarz and Minkowskii inequalities as given by (1.6.3) and (1.6.4) are

$$|(\phi, \psi)| \leqslant \|\phi\| \, \|\psi\| , \tag{8}$$

$$\|\phi + \psi\| \leqslant \|\phi\| + \|\psi\| . \tag{9}$$

Also recall the definition of an $\mathscr{L}_2$-kernel as given by (1.2.2)–(1.2.4).

Another concept that is fundamental in the theory of Hilbert spaces is the concept of completeness. A metric space is called complete if every Cauchy sequence of functions in this space is a convergent sequence. If a metric space is not complete, then there is a simple way to add elements to this space to make it complete. A Hilbert space is an inner-product linear space that is complete in its natural metric. The completeness of $\mathscr{L}_2$-spaces plays an important role in the theory of linear operators such as the Fredholm operator K,

$$K\phi = \int K(s,t)\phi(t)\,dt\ .\tag{10}$$

The operator adjoint to K is

$$K^*\psi = \int K^*(t,s)\psi(t)\,dt\ .\tag{11}$$

The operators (10) and (11) are connected by the interesting relationship

$$(K\phi,\psi) = (\phi,K^*\psi)\ ,\tag{12}$$

which is proved as follows:

$$\begin{aligned}
(K\phi,\psi) &= \int \psi^*(s)\left[\int K(s,t)\phi(t)\,dt\right]ds\\
&= \int \phi(t)\left[\int K(s,t)\psi^*(s)\,ds\right]dt\\
&= \int \phi(s)\left[\int K(t,s)\psi^*(t)\,dt\right]ds\\
&= \int \phi(s)\left[\int K^*(t,s)\psi(t)\,dt\right]^*ds\\
&= (\phi,K^*\psi)\ .
\end{aligned}$$

For a symmetric kernel, this result becomes

$$(K\phi,\psi) = (\phi,K\psi)\ ,\tag{13}$$

i.e., a symmetric operator is self-adjoint. Note that permutation of factors in a scalar product is equivalent to taking the complex conjugate, i.e., $(\phi,K\phi) = (K\phi,\phi)^*$. Combining this with (13), we find that, *for a symmetric kernel, the inner product $(K\phi,\phi)$ is always real*; the converse is also true and forms Exercise 1 at the end of this chapter.

An Orthonormal System of Functions

Systems of orthogonal functions play an important role in the theory of integral equations and their applications. A finite or an infinite set $\{\phi_k\}$ is said to be an orthogonal set if $(\phi_i, \phi_j) = 0$, $i \neq j$. If none of the elements of this set is a zero vector, then it is said to be a proper orthogonal set. A set is orthonormal if

$$(\phi_i, \phi_j) = \begin{cases} 0, & i \neq j, \\ 1, & i = j. \end{cases}$$

Individual functions ϕ for which $\|\phi\| = 1$ are said to be normalized.

Given a finite or an infinite (denumerable) independent set of functions $\{\psi_1, \psi_2, ..., \psi_k, ...\}$, we can construct an orthonormal set $\{\phi_1, \phi_2, ..., \phi_k, ...\}$ by the well-known Gram–Schmidt procedure as follows.

Let $\phi_1 = \psi_1 / \|\psi_1\|$. To obtain ϕ_2, we define

$$w_2(s) = \psi_2(s) - (\psi_2, \phi_1)\phi_1 .$$

The function w_2 is clearly orthogonal to ϕ_1; thus, ϕ_2 is obtained by setting $\phi_2 = w_2 / \|w_2\|$. Continuing this process, we have

$$w_k(s) = \psi_k(s) - \sum_{i=1}^{k-1} (\psi_k, \phi_i)\phi_i , \qquad \phi_k = w_k / \|w_k\| .$$

We have, thereby, obtained an equally numerous set of orthonormal functions. In case we are given a set of orthogonal functions, we can convert it into an orthonormal set simply by dividing each function by its norm.

Starting from an arbitrary orthonormal system, it is possible to construct the theory of Fourier series, analogous to the theory of trigonometric series. Suppose we want to find the best approximation of an arbitrary function $\psi(x)$ in terms of a linear combination of an orthonormal set $(\phi_1, \phi_2, ..., \phi_n)$. By the best approximation, we mean that we choose the coefficients $\alpha_1, \alpha_2, ..., \alpha_n$ such as to minimize $\|\psi - \sum_{i=1}^{n} \alpha_i \phi_i\|$, or, what is equivalent, to minimize $\|\psi - \sum_{i=1}^{n} \alpha_i \phi_i\|^2$.

Now, for any $\alpha_1, ..., \alpha_n$, we have

$$\|\psi - \sum_{i=1}^{n} \alpha_i \phi_i\|^2 = \|\psi\|^2 + \sum_{i=1}^{n} |(\psi, \phi_i) - \alpha_i|^2 - \sum_{i=1}^{n} |(\psi, \phi_i)|^2 . \qquad (14)$$

It is obvious that the minimum is achieved by choosing $\alpha_i = (\psi, \phi_i) = a_i$ (say). The numbers a_i are called the Fourier coefficients of the function $\psi(s)$ relative to the orthonormal system $\{\phi_i\}$. In that case, the relation (14) reduces to

$$\|\psi - \sum_{i=1}^{n} \alpha_i \phi_i\|^2 = \|\psi\|^2 - \sum_{i=1}^{n} |a_i|^2 . \qquad (15)$$

Since the left side is nonnegative, we have

$$\sum_{i=1}^{n} |a_i|^2 \leqslant \|\psi\|^2 , \qquad (16)$$

which, for the infinite set $\{\phi_i\}$, leads to the *Bessel inequality*

$$\sum_{i=1}^{\infty} |a_i|^2 \leqslant \|\psi\|^2 . \qquad (17)$$

Suppose we are given an infinite orthonormal system $\{\phi_i(s)\}$ in $\mathscr{L}_2$, and a sequence of constants $\{\alpha_i\}$, then the convergence of the series $\sum_{k=1}^{\infty} |\alpha_k^2|$ is evidently a necessary condition for the existence of an $\mathscr{L}_2$-function $f(s)$ whose Fourier coefficients with respect to the system ϕ_i are α_i. It so happens that this condition is also sufficient and the result is embodied in the Riesz–Fischer theorem, which we state as follows without proof.

Riesz–Fischer Theorem. If $\{\phi_i(s)\}$ is a given orthonormal system of functions in $\mathscr{L}_2$ and $\{\alpha_i\}$ is a given sequence of complex numbers such that the series $\sum_{k=1}^{\infty} |\alpha_k^2|$ converges, then there exists a unique function $f(s)$ for which α_i are the Fourier coefficients with respect to the orthonormal system $\{\phi_i\}$ and to which the Fourier series converges in the mean, that is,

$$\|f(s) - \sum_{i=1}^{n} \alpha_i \phi_i\| \to 0 \qquad \text{as} \quad n \to \infty .$$

If an orthonormal system of functions ϕ_i can be found in $\mathscr{L}_2$-space such that every other element of this space can be represented linearly in terms of this system, then it is called an orthonormal basis. The concepts of an orthonormal basis and a complete system of orthonormal functions are equivalent. Indeed, if any of the following criteria are met, the orthonormal set $\{\phi_1, ..., \phi_k, ...\}$ is complete.

(a)   For every function ψ in $\mathscr{L}_2$,

$$\psi = \sum_i (\psi, \phi_i) \phi_i = \sum_i a_i \phi_i .$$

(18)

(b)   For every function ψ in $\mathscr{L}_2$,

$$\|\psi\|^2 = \sum_{i=1}^{\infty} |(\psi, \phi_i)|^2 .$$

(19)

This is called *Parseval's identity*.

(c)   The only function ψ in $\mathscr{L}_2$ for which all the Fourier coefficients vanish is the trivial function (zero function).

(d)   There exists no function ψ in $\mathscr{L}_2$ such that $\{\psi, \phi_1, ..., \phi_k, ...\}$ is an orthonormal set.

The equivalence of these different criteria can be easily proved.

One frequently encounters Fourier series of somewhat more general character. Let $r(t)$ be a continuous, real, and nonnegative function in the interval (a, b). We say that the set of functions $\{\phi_i\}$ is orthonormal with weight $r(t)$ if

$$\int r(t) \phi_j(t) \phi_k(t) \, dt = \begin{cases} 0, & j \neq k , \\ 1, & j = k . \end{cases}$$

(20)

The Fourier expansions in terms of such functions are treated by introducing a new inner product

$$(\phi, \psi) = \int r(t) \phi(t) \psi^*(t) \, dt$$

(21)

with the corresponding norm

$$\|\phi\|_r = \left[\int r(t) \phi(t) \phi^*(t) \, dt \right]^{\frac{1}{2}} .$$

(22)

The space of functions for which $\|\phi\|_r < \infty$ is a Hilbert space and all the above results hold.

Some examples of the complete orthogonal and orthonormal systems are listed below.

(a)   The system $\phi_k(s) = (2\pi)^{-\frac{1}{2}} e^{iks}$ is orthonormal, where k is any integer $-\infty < k < \infty$.

(b)   The Legendre polynomials

$$P_0(s) = 1, \qquad P_n(s) = \frac{1}{2^n n!} \frac{d^n (s^2 - 1)^n}{ds^n}, \qquad n = 1, 2, \ldots$$

are orthogonal in the interval $(-1, 1)$. Indeed,

$$\int_{-1}^{1} P_j(s) P_k(s) \, ds = \begin{cases} 0, & j \neq k, \\ 2/(2k+1), & j = k. \end{cases} \tag{23}$$

(c)   Let $\alpha_{k,n}$ denote the positive zeros of the Bessel function $J_n(s)$, $k = 1, 2, \ldots$; $n > -1$. The system of functions $J_n(\alpha_{k,n} s)$ is orthogonal with weight $r(s) = s$ in the interval $(0, 1)$:

$$\int_{0}^{1} s J_n(\alpha_{j,n} s) J_n(\alpha_{k,n} s) \, ds = \begin{cases} 0, & j \neq k, \\ J_{n+1}^2 (\alpha_{k,n}), & j = k. \end{cases} \tag{24}$$

A Complete Two-Dimensional Orthonormal Set over the Rectangle $a \leqslant s \leqslant b, c \leqslant t \leqslant d$

Let $\{\phi_i(s)\}$ be a complete orthonormal set over $a \leqslant s \leqslant b$, and let $\{\psi_i(t)\}$ be a complete orthonormal set over $c \leqslant t \leqslant d$. Then, we claim that the set $\phi_1(s)\psi_1(t), \phi_1(s)\psi_2(t), \ldots, \phi_2(s)\psi_1(t), \ldots$ is a complete two-dimensional orthonormal set over the rectangle $a \leqslant s \leqslant b, c \leqslant t \leqslant d$.

The fact that the sequence of two-dimensional functions $\{\phi_i(s)\psi_j(t)\}$ is an orthonormal set follows readily by integrating over the rectangle. The completeness is proved by showing that every continuous function $F(s, t)$ with finite norm $\|F\|$, whose Fourier coefficients with respect to the set $\{\phi_i(s)\psi_j(t)\}$ are all zero vanishes identically over the rectangle.

For details on the results and ideas of this entire section, the reader is referred to [3, 19].

7.2. FUNDAMENTAL PROPERTIES OF EIGENVALUES AND EIGENFUNCTIONS FOR SYMMETRIC KERNELS

We have discussed the eigenvalues and eigenfunctions for integral equations in the previous chapters. In Chapters 2 and 4, we found that the eigenvalues of an integral equation are the zeros of certain determinants. It may happen that no such zeros exist, so that the kernel has no eigenvalues. There are many kernels for which there are no eigenvalues. Consider, for example, the homogeneous equation (cf. Exercise 3, Chapter 2)

$$g(s) = \lambda \int_0^{2\pi} (\sin s \cos t) g(t) \, dt = \lambda (\sin s) \int_0^{2\pi} g(t) \cos t \, dt . \tag{1}$$

Its solution must clearly be of the form $g(s) = A \sin s$. Substituting this in (1) yields

$$A \sin s = \lambda (\sin s) \int_0^{2\pi} A (\sin t \cos t) \, dt = 0 .$$

Thus, the kernel $K(s, t) = \sin s \cos t$ $(0 \leqslant s \leqslant 2\pi, \ 0 \leqslant t \leqslant 2\pi)$ has no eigenvalues. For a symmetric kernel that is nonnull (i.e., that is not identically zero), the above phenomenon cannot occur. Indeed, a *symmetric kernel possesses at least one eigenvalue*. The proof is briefly discussed in Section 7.8.

An eigenvalue is simple if there is only one corresponding eigenfunction, otherwise the eigenvalues are degenerate. The spectrum of the kernel K is the set of all its eigenvalues. In this terminology, the above assumption states that *the spectrum of a symmetric kernel is never empty*.

The following are some important properties of the symmetric integral equations

$$\lambda \int K(s, t) g(t) \, dt = f(s), \quad \text{or} \quad \lambda K g = f ; \quad K(s, t) = K^*(t, s) . \tag{2}$$

1. The eigenvalues of a symmetric kernel are real. Let λ and $\phi(s)$ be an eigenvalue and a corresponding eigenfunction of the kernel $K(s, t)$. This means

$$\phi(s) - \lambda K \phi(s) = 0 . \tag{3}$$

Multiply (3) by $\phi^*(s)$ and integrate with respect to s from a to b and derive the relation

$$\|\phi(s)\|^2 - \lambda (K\phi, \phi) = 0$$

or

$$\lambda = \|\phi(s)\|^2 / (K\phi, \phi) . \tag{4}$$

Since both the numerator and denominator are real, we have the required result.

2. The eigenfunctions of a symmetric kernel, corresponding to different eigenvalues, are orthogonal. If ϕ_1 and ϕ_2 are eigenfunctions corresponding, respectively, to the eigenvalues λ_1 and λ_2, we have

$$\phi_1 - \lambda_1 K\phi_1 = 0, \qquad \phi_2 - \lambda_2 K\phi_2 = 0. \tag{5}$$

Since λ_2 is real, the second equation in (5) may be written as $\phi_2{}^* - \lambda_2 K^* \phi_2{}^* = 0$. Then, by suitable multiplication and integration it follows that

$$\lambda_1 \lambda_2 \left[\iint \phi_2{}^*(s) K(s,t) \phi_1(t)\, dt\, ds \right.$$
$$\left. - \iint \phi_1(s) K^*(s,t) \phi_2{}^*(t)\, dt\, ds \right] = (\lambda_2 - \lambda_1)(\phi_1, \phi_2). \tag{6}$$

But $K(s,t) = K^*(t,s)$, and the left side vanishes identically, and because $\lambda_1 \neq \lambda_2$, the right side becomes $(\phi_1, \phi_2) = 0$.

3. The multiplicity of any nonzero eigenvalue is finite for every symmetric kernel for which $\iint |K(s,t)|^2\, ds\, dt$ is finite. Let the functions $\phi_{1\lambda}(s), \phi_{2\lambda}(s), \ldots, \phi_{n\lambda}(s) \ldots$ be the linearly independent eigenfunctions which correspond to a nonzero eigenvalue λ. By the Gram–Schmidt procedure, we can find linear combinations of these functions which form an orthonormal system $\{u_{k\lambda}(s)\}$. Then, the corresponding complex conjugate system $\{u_{k\lambda}{}^*\}$ also forms an orthonormal system. Let

$$K(s,t) \sim \sum_i a_i u_{i\lambda}{}^*(t),$$

$$a_i = \int K(s,t) u_{i\lambda}(t)\, dt = \lambda^{-1} u_{i\lambda}(s)$$

be the series associated with the kernel $K(s,t)$ for a fixed s. Applying Bessel's inequality to this series, we have

$$\int |K(s,t)|^2\, dt \geqslant \sum_i (\lambda^{-1})^2 |u_{i\lambda}(s)|^2,$$

which, when integrated with respect to s, yields

$$\iint |K(s,t)|^2\, ds\, dt \geqslant \sum_i (\lambda^{-1})^2. \tag{7}$$

The right side of (7) is $m(\lambda^{-1})^2$, where m is the multiplicity of λ. Since the left side is finite, it follows that m is finite.

4. The sequence of eigenfunctions of a symmetric kernel can be made orthonormal. Suppose that, corresponding to a certain eigenvalue, there are m linearly independent eigenfunctions. In view of the linearity of the integral operator, every linear combination of these functions is also an eigenfunction. Thus, by the Gram–Schmidt procedure, we can get equally numerous eigenfunctions which are orthonormal.

On the other hand, for different eigenvalues, the corresponding eigenfunctions are orthogonal and can be readily normalized. Combining these two facts, we have the proof of the above property.

5. The eigenvalues of a symmetric $\mathcal{L}_2$-kernel form a finite or an infinite sequence $\{\lambda_n\}$ with no finite limit point. If we include each eigenvalue in the sequence a number of times equal to its multiplicity, then

$$\sum_{n=1}^{\infty} (\lambda_n^{-1})^2 \leqslant \iint |K(s,t)|^2 \, ds \, dt \, . \tag{8}$$

Let $\{u_k(s)\}$ be the orthonormal eigenfunctions corresponding to different (nonzero) eigenvalues λ_i. Then, proceeding as in the proof of the property 3 and applying the Bessel inequality, we have

$$\sum_i (\lambda_i^{-1})^2 \leqslant \iint |K(s,t)|^2 \, ds \, dt < \infty \, .$$

Hence, if there exists an enumerable infinity of λ_i, then we must have $\sum_{n=1}^{\infty} (\lambda_n^{-1})^2 < \infty$. It follows that $\lim(1/\lambda_i) \to 0$ and ∞ is the only limit point of the eigenvalues.

6. The set of eigenvalues of the second iterated kernel coincide with the set of squares of the eigenvalues of the given kernel.

Note that the symmetry of the kernel shall not be assumed to prove this result.

Let λ be an eigenvalue of K with corresponding eigenfunction $\phi(s)$, that is, $(I - \lambda K)\phi = 0$, where I is the identity operator. When we operate on both sides of this equation with the operator $(I + \lambda K)$, we obtain $(I - \lambda^2 K^2)\phi = 0$ or

$$\phi(s) - \lambda^2 \int K_2(s,t)\phi(t) \, dt = 0 \, , \tag{9}$$

which proves that λ^2 is an eigenvalue of the kernel $K_2(s,t)$.

Conversely, let $\mu = \lambda^2$ be an eigenvalue of the kernel $K_2(s, t)$, with $\phi(s)$ as the corresponding eigenfunction. Then, $(I - \lambda^2 K^2)\phi = 0$ or

$$(I - \lambda K)(I + \lambda K)\phi = 0 . \tag{10}$$

If λ is an eigenvalue of K, then the above property is proved. If not, let us set $(I + \lambda K)\phi = \phi'(s)$ in (10) and obtain $(I - \lambda K)\phi'(s) = 0$. Since we have assumed that λ is not an eigenvalue of K, it follows that $\phi'(s) \equiv 0$, or equivalently $(1 + \lambda K)\phi = 0$. Thus, $-\lambda$ is an eigenvalue of the kernel K and the above property is proved.

We can extend the above result to the nth iterate. The set of eigenvalues of the kernel $K_n(s, t)$ coincide with the set of nth powers of the eigenvalues of the kernel $K(s, t)$.

7. If λ_1 is the smallest eigenvalue of the kernel K, then

$$1/|\lambda_1| = \max\left[|(K\phi, \phi)|/\|\phi\|\right]$$

$$= \max\left[|\iint K(s, t)\phi(t)\phi^*(s)\, dt\, ds|/\|\phi\|\right] \tag{11}$$

or equivalently,

$$1/|\lambda_1| = \max(K\phi, \phi), \qquad \|\phi\| = 1 . \tag{12}$$

This maximum value is attained when $\phi(s)$ is an eigenfunction of the symmetric $\mathscr{L}_2$-kernel corresponding to the smallest eigenvalue. For proof, see Section 7.8.

7.3. EXPANSION IN EIGENFUNCTIONS AND BILINEAR FORM

We now discuss the results concerning the expansion of a symmetric kernel and of functions represented in a certain sense by the kernel, in terms of its eigenfunctions and the eigenvalues. Recall that we meet a similar situation when we deal with a Hermitian matrix. For instance, if A is a Hermitian matrix, then there is a unitary matrix U such that $U^{-1}AU$ is diagonal. This means that, by transforming to an orthonormal basis of the vector space consisting of the eigenfunctions of A, the matrix representing the operator A becomes diagonal.

Let $K(s, t)$ be a nonnull, symmetric kernel which has a finite or an infinite number of eigenvalues (always real and nonzero). We order them in the sequence

$$\lambda_1, \lambda_2, ..., \lambda_n, ... \tag{1}$$

in such a way that each eigenvalue is repeated as many times as its multiplicity. We further agree to denumerate these eigenvalues in the order that corresponds to their absolute values, i.e.,

$$0 < |\lambda_1| \leqslant |\lambda_2| \leqslant \cdots \leqslant |\lambda_n| \leqslant |\lambda_{n+1}| \leqslant \cdots .$$

Let

$$\phi_1(s), \phi_2(s), ..., \phi_n(s), ... \tag{2}$$

be the sequence of eigenfunctions corresponding to the eigenvalues given by the sequence (1) and arranged in such a way that they are no longer repeated and are linearly independent in each group corresponding to the same eigenvalue. Thus, to each eigenvalue λ_k in (1) there corresponds just one eigenfunction $\phi_k(s)$ in (2). According to the property 4 of the previous section, we assume that they have been orthonormalized.

Now, we have assumed that a symmetric $\mathscr{L}_2$-kernel has at least one eigenvalue, say λ_1. Then $\phi_1(s)$ is the corresponding eigenfunction. It follows that the "truncated" symmetric kernel

$$K^{(2)}(s, t) = K(s, t) - [\phi_1(s)\phi_1{}^*(t)/\lambda_1]$$

is nonnull and it will also have at least one eigenvalue λ_2 (we choose the smallest if there are more) with corresponding normalized eigenfunction $\phi_2(s)$. The function $\phi_1(s) \neq \phi_2(s)$ even if $\lambda_1 = \lambda_2$, since

$$\int K^{(2)}(s, t)\phi_1(t)\, dt = \int K(s, t)\phi_1(t)\, dt$$

$$- \frac{\phi_1(s)}{\lambda_1} \int \phi_1(t)\phi_1{}^*(t)\, dt$$

$$= \frac{\phi_1(s)}{\lambda_1} - \frac{\phi_1(s)}{\lambda_1} = 0 .$$

Similarly, the third truncated kernel

$$K^{(3)}(s, t) = K^{(2)}(s, t) - \frac{\phi_2(s)\phi_2{}^*(t)}{\lambda_2}$$

$$= K(s, t) - \sum_{k=1}^{2} \frac{\phi_k(s)\phi_k{}^*(t)}{\lambda_k}$$

gives the third eigenvalue λ_3 and eigenfunction $\phi_3(s)$. Continuing in this way, we end up with two possibilities: either this process terminates after n steps, that is, $K^{(n+1)}(s, t) \equiv 0$, and the kernel $K(s, t)$ is a degenerate kernel,

$$K(s, t) = \sum_{k=1}^{n} \frac{\phi_k(s)\,\phi_k{}^*(t)}{\lambda_k}, \tag{3}$$

or the process can be continued indefinitely and there are an infinite number of eigenvalues and eigenfunctions.

Note that we have denoted the least eigenvalue and the corresponding eigenfunction of $K^{(n)}(s, t)$ as λ_n and ϕ_n, which are the nth eigenvalue and the nth eigenfunction in the sequences (1) and (2). This will be justified by Theorem 2 below.

We shall examine in the next section whether the bilinear form (3) is valid for the case when the kernel $K(s, t)$ has infinite eigenvalues and eigenfunctions. The following two theorems, however, follow readily.

Theorem 1.   Let the sequence $\{\phi_k(s)\}$ be all the eigenfunctions of a symmetric $\mathscr{L}_2$-kernel $K(s, t)$, with $\{\lambda_k\}$ as the corresponding eigenvalues. Then, the series

$$\sum_{n=1}^{\infty} \frac{|\phi_n(s)|^2}{\lambda_n{}^2}$$

converges and its sum is bounded by $C_1{}^2$, which is an upper bound of the integral

$$\int |K^2(s, t)|\, dt .$$

Proof.   This result is an immediate consequence of Bessel's inequality. Indeed, the Fourier coefficients a_n of the function $K(s, t)$, with fixed s, with respect to the orthonormal system $\{\phi_n{}^*(s)\}$ are

$$a_n = \int K(s, t)\,\phi_n(t)\, dt = \phi_n(s)/\lambda_n .$$

Thus, applying Bessel's inequality, we have

$$\sum_{n=1}^{\infty} \frac{|\phi_n(s)|^2}{\lambda_n{}^2} \leqslant \int |K(s, t)|^2\, dt \leqslant C_1{}^2 . \tag{4}$$

Theorem 2. Let the sequence $\phi_n(s)$ be all the eigenfunctions of a symmetric kernel $K(s, t)$, with $\{\lambda_n\}$ as the corresponding eigenvalues. Then, the truncated kernel

$$K^{(n+1)}(s, t) = K(s, t) - \sum_{m=1}^{n} \frac{\phi_m(s)\,\phi_m{}^*(t)}{\lambda_m} \tag{5}$$

has the eigenvalues $\lambda_{n+1}, \lambda_{n+2}, \dots$, to which correspond the eigenfunctions $\phi_{n+1}(s), \phi_{n+2}(s), \dots$. The kernel $K^{(n+1)}(s, t)$ has no other eigenvalues or eigenfunctions.

Proof. (a) Observe that the integral equation

$$\phi(s) - \lambda \int K^{(n+1)}(s, t)\,\phi(t)\,dt = 0 \tag{6}$$

is equivalent to

$$\phi(s) - \lambda \int K(s, t)\,\phi(t)\,dt + \lambda \sum_{m=1}^{n} \frac{\phi_m(s)}{\lambda_m}\,(\phi, \phi_m)\,dt = 0 \;. \tag{7}$$

If, on the left side of this equation we set $\lambda = \lambda_j$ and $\phi(s) = \phi_j(s)$ $j \geqslant n+1$, then, in view of the orthogonality condition, we have

$$\phi_j(s) - \lambda_j \int K(s, t)\,\phi_j(t)\,dt = 0 \;. \tag{8}$$

This means that ϕ_j and λ_j for $j \geqslant n+1$ are the eigenfunctions and eigenvalues of the kernel $K^{(n+1)}(s, t)$.

(b) Let λ and $\phi(s)$ be an eigenvalue and eigenfunction of the kernel $K^{(n+1)}(s, t)$ so that

$$\phi(s) - \lambda K\phi(s) + \lambda \sum_{m=1}^{n} \frac{\phi_m(s)}{\lambda_m}\,(\phi, \phi_m) = 0 \;. \tag{9}$$

Taking the scalar product of (9) with $\phi_j(s), j \leqslant n$, we obtain

$$(\phi, \phi_j) - \lambda(K\phi, \phi_j) + (\lambda/\lambda_j)(\phi, \phi_j) = 0 \;, \tag{10}$$

where we have used the orthonormality of the ϕ_j. But $(K\phi, \phi_j) = (\phi, K\phi_j)$ $= \lambda_j{}^{-1}(\phi, \phi_j)$. Hence, (10) becomes

$$(\phi, \phi_j) + (\lambda/\lambda_j)\,\{(\phi, \phi_j) - (\phi, \phi_j)\} = (\phi, \phi_j) = 0 \;. \tag{11}$$

Thereby, the last term in the left side of equation (9) vanishes and we are left with

$$\phi(s) - \lambda \int K(s,t)\phi(t)\, dt = 0 , \tag{12}$$

which means that λ and $\phi(s)$ are eigenvalue and eigenfunction of the kernel $K(s,t)$ and that $\phi \neq \phi_j, j \leqslant n$. Indeed, ϕ is orthogonal to all ϕ_j, $j \leqslant n$, and $\phi(s)$ and λ are necessarily contained in the sequences $\{\phi_k(s)\}$ and $\{\lambda_k\}$, $k \geqslant n+1$, respectively.

In light of the above two theorems, we can easily conclude that, if the symmetric kernel K has only a finite number of eigenvalues, then it is degenerate. The proof follows by observing that $K^{(n+1)}(s,t)$ then has no eigenvalues and hence it must be null. Therefore,

$$K(s,t) = \sum_{m=1}^{n} \frac{\phi_m(s)\,\phi_m^*(t)}{\lambda_m} .$$

In Chapter 2, we found that every degenerate kernel has only a finite number of eigenvalues. Combining these two results, we have the following theorem.

Theorem 3. A necessary and sufficient condition for a symmetric $\mathscr{L}_2$-kernel to be degenerate is that it have a finite number of eigenvalues.

7.4. HILBERT–SCHMIDT THEOREM AND SOME IMMEDIATE CONSEQUENCES

The pivotal result in the theory of symmetric integral equations is embodied in the following theorem.

Hilbert–Schmidt Theorem. If $f(s)$ can be written in the form

$$f(s) = \int K(s,t)h(t)\, dt , \tag{1}$$

where $K(s,t)$ is a symmetric $\mathscr{L}_2$-kernel and $h(t)$ is an $\mathscr{L}_2$-function, then $f(s)$ can be expanded in an absolutely and uniformly convergent Fourier series with respect to the orthonormal system of eigenfunctions (7.3.2) of the kernel K:

$$f(s) = \sum_{n=1}^{\infty} f_n \phi_n(s), \qquad f_n = (f, \phi_n).$$

The Fourier coefficients of the function $f(s)$ are related to the Fourier coefficients h_n of the function $h(s)$ by the relations

$$f_n = h_n/\lambda_n, \qquad h_n = (h, \phi_n), \tag{2}$$

where λ_n are the eigenvalues (7.3.1) of the kernel K.

Proof.   The Fourier coefficients of the function $f(s)$ with respect to the orthonormal system $\{\phi_n(s)\}$ are

$$f_n = (f, \phi_n) = (Kh, \phi_n) = (h, K\phi_n) = \lambda_n^{-1}(h, \phi_n) = \lambda_n^{-1} h_n,$$

where we have used the self-adjoint property of the operator as well as the relation $\lambda_n K\phi_n = \phi_n$. Thus, the Fourier series for $f(s)$ is

$$f(s) \sim \sum_{n=1}^{\infty} f_n \phi_n(s) = \sum_{n=1}^{\infty} \frac{h_n}{\lambda_n} \phi_n(s). \tag{3}$$

The remainder term for this series can be estimated as follows:

$$\left| \sum_{k=n+1}^{n+p} h_k \frac{\phi_k(s)}{\lambda_k} \right|^2 \leq \sum_{k=n+1}^{n+p} h_k^2 \sum_{k=n+1}^{n+p} \frac{|\phi_k(s)|^2}{\lambda_k^2}$$

$$\leq \sum_{k=n+1}^{n+p} h_k^2 \sum_{k=1}^{\infty} \frac{\phi_k^2(s)}{\lambda_k^2}. \tag{4}$$

From the relation (7.3.4), we find that the above series is bounded. Also, because $h(s)$ is an $\mathscr{L}_2$-function, the series $\sum_{k=1}^{\infty} h_k^2$ is convergent and the partial sum $\sum_{k=n+1}^{n+p} h_k^2$ can be made arbitrarily small. Therefore, the series (3) converges absolutely and uniformly.

It remains to be shown that the series (3) converges to $f(s)$ in the mean. To this end, let us denote its partial sum as

$$\psi_n(s) = \sum_{m=1}^{n} \frac{h_m}{\lambda_m} \phi_m(s) \tag{5}$$

and estimate the value of $\|f(s) - \psi_n(s)\|$. Now,

$$f(s) - \psi_n(s) = Kh - \sum_{m=1}^{n} \frac{h_m}{\lambda_m} \phi_m(s) \tag{6}$$

(continued)

$$= Kh - \sum_{m=1}^{n} \frac{(h, \phi_m)}{\lambda_m} \phi_m(s) = K^{(n+1)} h , \qquad (6)$$

where $K^{(n+1)}$ is the truncated kernel as defined in the previous section. From (6), we obtain

$$\|f(s) - \psi_n(s)\|^2 = \|K^{(n+1)}h\|^2 = (K^{(n+1)}h, K^{(n+1)}h)$$
$$= (h, K^{(n+1)} K^{(n+1)} h) = (h, K_2^{(n+1)} h) , \qquad (7)$$

where we have used the self-adjointness of the kernel $K^{(n+1)}$ and also the relation $K^{(n+1)} K^{(n+1)} = K_2^{(n+1)}$. If we use property 6 of Section 7.2 and Theorem 2 of Section 7.3, we find that the least eigenvalue of the kernel $K_2^{(n+1)}$ is equal to λ_{n+1}^2. Furthermore, according to property 7 of Section 7.2, we have

$$1/\lambda_{n+1}^2 = \max[(h, K_2^{(n+1)} h)/(h, h)] , \qquad (8)$$

where we have omitted the modulus sign from the scalar product $(h, K_2^{(n+1)} h)$, as it is a positive quantity. Combining (7) and (8), we have

$$\|f(s) - \psi_n(s)\|^2 = (h, K_2^{(n+1)} h) \leqslant (h, h)/\lambda_{n+1}^2.$$

Since $\lambda_{n+1} \to \infty$, we find that $\|f(s) - \psi_n(s)\| \to 0$ as $n \to \infty$.

Finally, we use the relation

$$\|f - \psi\| \leqslant \|f - \psi_n\| + \|\psi_n - \psi\| , \qquad (9)$$

where ψ is the limit of the series with partial sum ψ_n, to prove that $f = \psi$. The first term on the right side of (9) tends to zero, as proved above. To prove that the second term also tends to zero, we observe that, since the series (3) converges uniformly, we have, for an arbitrarily small and positive ε,

$$|\psi_n(s) - \psi(s)| < \varepsilon ,$$

when n is sufficiently large. Hence, $\|\psi_n(s) - \psi(s)\| < \varepsilon(b - a)^{1/2}$ and the result follows.

Remark. Note that we assumed neither the convergence of the Fourier series $h(s)$ nor the completeness of the orthonormal system. We have merely used the fact that h is an $\mathscr{L}_2$-function.

An immediate consequence of the Hilbert–Schmidt theorem is the bilinear form of the type (7.3.3). Indeed, by definition,

$$K_m(s,t) = \int K(s,x) K_{m-1}(x,t)\, dx\,, \qquad m = 2,3,\ldots\,, \qquad (10)$$

which is of the form (1) with $h(s) = K_{m-1}(s,t)$; t fixed. The Fourier coefficient $a_k(t)$ of $K_m(s,t)$ with respect to the system of eigenfunctions $\{\phi_k(s)\}$ of $K(s,t)$ is

$$a_k(t) = \int K_m(s,t)\,\phi_k^*(s)\, ds = \lambda_k^{-m}\,\phi_k^*(t)\,.$$

It follows from the above theorem that all the iterated kernels $K_m(s,t)$, $m \geqslant 2$, of a symmetric $\mathscr{L}_2$-kernel can be represented by the absolutely and uniformly convergent series

$$K_m(s,t) = \sum_{k=1}^{\infty} \lambda_k^{-m}\,\phi_k(s)\,\phi_k^*(t)\,. \qquad (11)$$

By setting $s = t$ in (11) and integrating from a to b, we obtain

$$\sum_{k=1}^{\infty} \lambda_k^{-m} = \int K_m(s,s)\, ds = A_m\,, \qquad (12)$$

where A_m is the trace of the iterated kernel K_m.

Next, we apply the Riesz–Fischer theorem and find from (12) with $m = 2$ that the series

$$\sum_{k=1}^{\infty} \frac{\phi_k(s)\,\phi_k^*(t)}{\lambda_k} \qquad (13)$$

converges in the mean to a symmetric $\mathscr{L}_2$-kernel $K(s,t)$, which, considered as a Fredholm kernel, has precisely the sequence of numbers $\{\lambda_k\}$ as eigenvalues.

Definite Kernels and Mercer's Theorem

A symmetric $\mathscr{L}_2$-kernel K is said to be nonnegative-definite if $(K\phi, \phi) \geqslant 0$ for every $\mathscr{L}_2$-function ϕ; furthermore, K is positive-definite if, in addition, $(K\phi, \phi) = 0$ implies ϕ is null. The definitions of nonpositive-definite and negative-definite symmetric kernels follows in an obvious manner. A symmetric kernel that does not fall into any of these four categories is called indefinite.

The following theorem is an immediate consequence of the Hilbert–Schmidt theorem.

Theorem. A nonnull, symmetric $\mathscr{L}_2$-kernel K is nonnegative if and only if all its eigenvalues are positive; it is positive-definite if and only if the above condition is satisfied and, in addition, some (and therefore every) full orthonormal system of eigenfunctions of K is complete.

Proof. (a) From the Hilbert–Schmidt theorem, we have

$$f(s) = K\phi = \sum_{n=1}^{\infty} \frac{(\phi, \phi_n)}{\lambda_n} \phi_n(s) . \tag{14}$$

The result of taking the inner product of (14) with ϕ is

$$(K\phi, \phi) = \sum_{n=1}^{\infty} \frac{|(\phi, \phi_n)|^2}{\lambda_n} . \tag{15}$$

Hence, if all $\lambda_n > 0$, we have $(K\phi, \phi) \geqslant 0$ for all ϕ. In addition, if any λ_n is negative, then $(K\phi_n, \phi_n) = \lambda_n^{-1} < 0$. Thereby, the first part of the theorem is proved.

(b) Let K be nonnegative-definite. From (15), it follows that $(K\phi, \phi) = 0$ if and only if $(\phi, \phi_n) = 0$ for all n. Therefore, K will be positive-definite if and only if the vanishing of (ϕ, ϕ_n) for all n implies $\phi \equiv 0$. Using the criterion (d) for the completeness of an orthonormal system as given in Section 7.1, we find that the second part of the theorem is thereby proved.

Finally, we state without proof the following result, which gives the precise conditions for the bilinear form (7.3.3) to be extended to an infinite series.

Mercer's Theorem. If a nonnull, symmetric $\mathscr{L}_2$-kernel is quasi-definite (that is, when all but a finite number of eigenvalues are of one sign) and continuous, then the series

$$\sum_{n=1}^{\infty} \lambda_n^{-1} \tag{16}$$

is convergent and

$$K(s,t) = \sum_{n=1}^{\infty} \frac{\phi_n(s)\,\phi_n^*(t)}{\lambda_n} , \qquad (17)$$

the series being uniformly and absolutely convergent.

Note that the continuity of the kernel is an absolutely essential condition for the theorem to be true.

7.5. SOLUTION OF A SYMMETRIC INTEGRAL EQUATION

Let us use the Hilbert–Schmidt theorem to find an explicit solution of the inhomogeneous Fredholm integral equation of the second kind

$$g(s) = f(s) + \lambda \int K(s,t)\,g(t)\,dt \qquad (1)$$

with a symmetric $\mathscr{L}_2$-kernel. It is assumed that λ is not an eigenvalue and that all the eigenvalues and eigenfunctions of the kernel $K(s,t)$ are known and arranged as in (7.3.1) and (7.3.2), respectively. The first thing we observe is that the function $g(s) - f(s)$ has an integral representation of the form (7.4.1). As such, we can use the Hilbert–Schmidt theorem and write

$$g(s) - f(s) = \sum_{k=1}^{\infty} c_k \phi_k(s) , \qquad (2)$$

where

$$c_k = \int [g(s) - f(s)]\,\phi_k^*(s)\,ds = g_k - f_k \qquad (3)$$

with

$$g_k = \int g(s)\,\phi_k^*(s)\,ds , \qquad f_k = \int f(s)\,\phi_k^*(s)\,ds . \qquad (4)$$

Furthermore, the relation (7.4.2) gives

$$c_k = \lambda g_k/\lambda_k . \qquad (5)$$

Since λ is not an eigenvalue, from (3) and (5) we have

$$c_k = [\lambda/(\lambda_k - \lambda)]\,f_k , \qquad g_k = [\lambda_k/(\lambda_k - \lambda)]\,f_k . \qquad (6)$$

Substituting the value of c_k from (6) into (2), we derive the solution of the integral equation (1) in terms of an absolutely and uniformly convergent series:

$$g(s) = f(s) + \lambda \sum_{k=1}^{\infty} \frac{f_k}{\lambda_k - \lambda} \, \phi_k(s) \qquad (7)$$

or

$$g(s) = f(s) - \lambda \sum_{k=1}^{\infty} \int \frac{\phi_k(s) \, \phi_k^*(t)}{(\lambda - \lambda_k)} f(t) \, dt \; . \qquad (8)$$

Thus, the resolvent kernel $\Gamma(s, t; \lambda)$ can be expressed by the series

$$\Gamma(s, t; \lambda) = \sum_{k=1}^{\infty} \frac{\phi_k(s) \, \phi_k^*(t)}{(\lambda - \lambda_k)} \qquad (9)$$

from which it follows that, the singular points of the resolvent kernel Γ corresponding to a symmetric $\mathscr{L}_2$-kernel are simple poles and every pole is an eigenvalue of the kernel.

The above discussion is based on the assumption that λ is not an eigenvalue. If it is an eigenvalue, then it necessarily occurs in the sequence $\{\lambda_k\}$ and perhaps is repeated several times. Let $\lambda = \lambda_m = \lambda_{m+1} = \cdots = \lambda_{m'}$. For the indices k, different from $m, m+1, ..., m'$, the coefficients c_k

and g_k in (6) are well-defined. However, if k is equal to one of these numbers, then $f_k = 0$. This means that the integral equation (1) is soluble if, and only if, the function $f(s)$ is orthogonal to the eigenfunctions $\phi_m, \phi_{m+1}, ..., \phi_{m'}$. When this condition is satisfied, the solution is given by formula (7), where the coefficients with the indeterminate form $0/0$ have to be taken as arbitrary numbers.

Finally, we attempt to solve the Fredholm integral equation of the first kind

$$f(s) = \int K(s, t) g(t) \, dt \; , \qquad (10)$$

where the kernel $K(s, t)$ is a symmetric $\mathscr{L}_2$-kernel. We again assume that the sequence of eigenvalues $\{\lambda_k\}$ and corresponding eigenfunctions $\{\phi_k(s)\}$ are known and arranged as in (7.3.1) and (7.3.2).

From relation (7.4.2), we have

$$f_k = g_k / \lambda_k \qquad \text{or} \qquad g_k = \lambda_k f_k \; . \qquad (11)$$

Because of the Riesz–Fischer theorem, there are only two possibilities: either (a) the infinite series

$$\sum_{k=1}^{\infty} f_k^2 \, \lambda_k^2 \qquad (12)$$

diverges and equation (10) has no solution, or (b) the series (12) converges and there is a unique $\mathscr{L}_2$-function $g(s)$ which is the solution of equation (10). This solution can be evaluated by taking the limit in the mean

$$g(s) = \lim_{n \to \infty} \sum_{k=1}^{n} \lambda_k f_k \phi_k(s) .$$ (13)

7.6. EXAMPLES

Example 1.   Solve the symmetric integral equation

$$g(s) = (s+1)^2 + \int_{-1}^{1} (st + s^2 t^2) g(t) \, dt .$$ (1)

The eigenvalues and eigenfunctions for this symmetric kernel can be found by the method of Chapter 2 (see Example 4 in Section 2.2):

$$\begin{aligned} \lambda_1 &= \tfrac{3}{2}, & \phi_1(s) &= (\tfrac{1}{2}\sqrt{6})s , \\ \lambda_2 &= \tfrac{5}{2}, & \phi_2(s) &= [(1/\sqrt{2})\sqrt{10}]s^2 . \end{aligned}$$ (2)

$$\begin{aligned} f_1 &= \int_{-1}^{1} (t^2 + 2t + 1)(\tfrac{1}{2}\sqrt{6})t \, dt = \tfrac{2}{3}\sqrt{6} , \\ f_2 &= \int_{-1}^{1} (t^2 + 2t + 1)(\tfrac{1}{2}\sqrt{10})t^2 \, dt = (8/15)\sqrt{10} . \end{aligned}$$ (3)

Thus,

$$g(s) = \frac{(2/3)\sqrt{6}}{(3/2)-1}\left(\frac{1}{2}\sqrt{6}\right)s + \frac{(8/15)\sqrt{10}}{(5/2)-1}\left(\frac{1}{2}\sqrt{10}\right)s^2 + (s+1)^2$$

or

$$g(s) = (25/9)s^2 + 6s + 1 .$$ (4)

Note that, in equation (1), $\lambda = 1$, which is not an eigenvalue.

Example 2.   Solve the symmetric integral equation

$$g(s) = s^2 + 1 + \tfrac{3}{2} \int\limits_{-1}^{1} (st + s^2 t^2) g(s) \, ds \, . \qquad (5)$$

Here, $\lambda = \lambda_1 = \tfrac{3}{2}$, so we shall have the indeterminate form 0/0 in one of the coefficients. Fortunately, the function $(s^2 + 1)$ is orthogonal to the eigenfunction $(\tfrac{1}{2}\sqrt{6})s$, which corresponds to the eigenvalue $\tfrac{3}{2}$. Following the procedure of Section 7.5, we obtain

$$f_1 = 0, \quad f_2 = \int\limits_{-1}^{1} (t^2 + 1)(\tfrac{1}{2}\sqrt{10}) t^2 \, dt = (8/15)\sqrt{10} \, . \qquad (6)$$

Thus, the required solution is

$$g(s) = \frac{3}{2} \frac{(8/15)\sqrt{10}}{(5/2) - (3/2)} \left(\frac{1}{2}\sqrt{10} \right) s^2 + cs + s^2 + 1$$

or

$$g(s) = 5s^2 + cs + 1 \, , \qquad (7)$$

where c is an arbitrary constant.

Example 3.   Solve the symmetric integral equation

$$g(s) = f(s) + \lambda \int k(s) k(t) g(t) \, dt \, . \qquad (8)$$

If we write

$$\int k(s) k(t) k(t) \, dt = \left\{ \int [k(t)]^2 \, dt \right\} k(s) \, ,$$

we observe that

$$\lambda_1 = 1 \Big/ \int [k(t)]^2 \, dt \qquad (9)$$

is an eigenvalue. The corresponding normalized eigenfunction is

$$\phi_1(s) = k(s) \Big/ \left\{ \int [k(t)]^2 \, dt \right\}^{1/2} \, . \qquad (10)$$

The coefficient f_1 has the value

$$f_1 = \left\{ \int [k(t)]^2 \, dt \right\}^{-1/2} \int f(t) k(t) \, dt \, . \qquad (11)$$

Thus, for $\lambda \neq \lambda_1$, the solution is

$$g(s) = [\lambda f_1/(\lambda_1 - \lambda)]\,\phi_1(s) + f(s)$$

or

$$g(s) = \left(\lambda k(s)\int f(s)\,k(s)\,ds \Big/ \left\{1 - \lambda \int [k(s)]^2\,ds\right\}\right) + f(s)\,. \qquad (12)$$

On the other hand, if

$$\lambda = \lambda_1 = 1 \Big/ \int [k(s)]^2\,ds\,,$$

then $f(s)$ must be orthogonal to $\phi_1(s)$, and in that case the solution is

$$g(s) = f(s) + ck(s)\,, \qquad c \quad \text{an arbitrary constant}\,. \qquad (13)$$

Example 4. Solve the symmetric Fredholm integral equation of the first kind

$$\int\limits_0^1 K(s,t)\,g(t)\,dt = f(s) \qquad (14)$$

where

$$K(s,t) = \begin{cases} s(1-t)\,, & s < t\,, \\ (1-s)t\,, & s > t\,. \end{cases} \qquad (15)$$

Recall that in Example 2 of Section 5.3 we proved that the boundary value problem

$$\frac{d^2 y}{ds^2} + \lambda y = 0\,, \qquad y(0) = y(1) = 0 \qquad (16)$$

is equivalent to the homogeneous equation

$$g(s) = \lambda \int\limits_0^1 K(s,t)\,g(t)\,dt\,. \qquad (17)$$

The eigenvalues of the system (16) are

$$\lambda_1 = \pi^2\,, \qquad \lambda_2 = (2\pi)^2\,, \qquad \lambda_3 = (3\pi)^2\,, \qquad \dots\,,$$

and the corresponding normalized eigenfunctions are

$$\sqrt{2}\sin \pi s\,, \qquad \sqrt{2}\sin 2\pi s\,, \qquad \sqrt{2}\sin 3\pi s\,, \qquad \dots\,. \qquad (18)$$

Therefore,

$$f_k = \sqrt{2} \int_0^1 (\sin k\pi t) f(t) \, dt \, , \tag{19}$$

and the integral equation (14) has a solution of class $\mathscr{L}_2$ if and only if the infinite series

$$\sum_{k=1}^{\infty} f_k^2 \lambda_k^2 = \pi^4 \sum_{k=1}^{\infty} (k^4 f_k^2)$$

converges.

Example 5. Solve Poisson's integral equation

$$f(\theta) = \frac{1-\rho^2}{2\pi} \int_0^{2\pi} \frac{g(\alpha) \, d\alpha}{1 - 2\rho [\cos(\theta - \alpha)] + \rho^2} \, ,$$

$$0 \leqslant \theta \leqslant 2\pi \, ; \qquad 0 < \rho < 1 \, . \tag{20}$$

Here, the symmetric kernel $K(\theta, \alpha)$ can be expanded to give

$$K(\theta, \alpha) = [(1 - \rho^2)/2\pi] \{1 - 2\rho [\cos(\theta - \alpha)] + \rho^2\}^{-1}$$

$$= (1/2\pi) + (1/\pi) \sum_{k=1}^{\infty} \rho^k \cos[k(\theta - \alpha)] \, . \tag{21}$$

It is a matter of simple verification that, by using the expansion (21), one gets $\int_0^{2\pi} K(\theta, \alpha) \, d\alpha = 1$, or

$$\int_0^{2\pi} K(\theta, \alpha) (2\pi)^{-1/2} \, d\alpha = (2\pi)^{-1/2} \, ,$$

which means that $\lambda_0 = 1$, $\phi_0(s) = (2\pi)^{-1/2}$. Similarly, using the formula

$$\int_0^{2\pi} K(\theta, \alpha) \frac{\cos}{\sin} n\alpha \, d\alpha = \rho^n \frac{\cos}{\sin} n\theta \, , \qquad n = 1, 2, 3, \ldots \, ,$$

we have

$$\lambda_{2k-1} = \lambda_{2k} = \rho^{-k} \, ; \qquad \phi_{2k-1}(s) = \pi^{-1/2} \cos ks \, ;$$

$$\phi_{2k}(s) = \pi^{-1/2} \sin ks \, , \qquad k = 1, 2, 3, \ldots \, . \tag{22}$$

We can now readily evaluate the coefficients f_k in the series (7.5.12), and it emerges that the integral equation has an $\mathcal{L}_2$-solution if and only if the infinite series

$$\sum_{n=1}^{\infty} \frac{a_n^2 + b_n^2}{\rho^{2n}} ,$$

where

$$a_n = (1/\pi) \int_0^{2\pi} f(\theta) \cos n\theta \, d\theta , \qquad b_n = (1/\pi) \int_0^{2\pi} f(\theta) \sin n\theta \, d\theta \quad (23)$$

converge.

7.7. APPROXIMATION OF A GENERAL $\mathcal{L}_2$-KERNEL (NOT NECESSARILY SYMMETRIC) BY A SEPARABLE KERNEL

In Section 2.5, we approximated an analytic kernel $s(e^{st} - 1)$ by a separable kernel. In this section, we show that we can approximate every $\mathcal{L}_2$-kernel in the mean by a separable kernel. The proof rests on the availability of a two-dimensional complete orthonormal set as discussed at the end of Section 7.1.

Let $K(s, t)$ be an $\mathcal{L}_2$-kernel and let $\{\psi_i(s)\}$ be an arbitrary, complete, orthonormal set over $a \leqslant s \leqslant b$. Then, the set $\{\psi_i(s)\psi_j^*(t)\}$ is a complete orthonormal set over the square $a \leqslant s, t \leqslant b$. The Fourier expansion of the kernel $K(s, t)$ in this set is

$$K(s, t) = \sum_{i, j=1}^{n} K_{ij} \psi_i(s) \psi_j^*(t) , \tag{1}$$

where the K_{ij} are the Fourier coefficients

$$K_{ij} = \iint K(s, t) \psi_i^*(s) \psi_j(t) \, ds \, dt . \tag{2}$$

Parseval's identity gives

$$\iint |K(s, t)|^2 \, ds \, dt = \sum_{i, j=1}^{\infty} |K_{ij}|^2 . \tag{3}$$

Now, if we define a separable kernel $k(s, t)$ as

$$k(s,t) = \sum_{i,j=1}^{n} K_{ij} \psi_i(s) \psi_j^*(t) ,$$

we find after a simple calculation that

$$\iint |K(s,t) - k(s,t)|^2 \, ds \, dt = \sum_{i,j=n+1}^{\infty} |K_{ij}|^2 . \tag{4}$$

But the sum in (4) can be made as small as we desire by choosing a sufficiently large n because the series (3) is convergent by hypothesis. This proves our assertion.

7.8. THE OPERATOR METHOD IN THE THEORY OF INTEGRAL EQUATIONS

In this section, we shall show briefly how we can treat a Fredholm integral equation from the standpoint of present-day functional analysis. We have already analyzed the properties of a function space in Section 7.1. We now note that the transformation or operator K,

$$K\phi = \int K(s,t) \phi(t) \, dt , \tag{1}$$

is linear, inasmuch as

$$K(\phi_1 + \phi_2) = K\phi_1 + K\phi_2 , \qquad K(\alpha\phi) = \alpha K\phi .$$

The operator K is called bounded if $\|K\phi\| \leqslant M \|\phi\|$ for an $\mathscr{L}_2$-kernel $K(s,t)$, an $\mathscr{L}_2$-function ϕ, and a constant M. The norm $\|K\|$ of K is defined as

$$\|K\| = \text{l.u.b.} (\|Kg\|/\|g\|) ,$$

or

$$\|K\| = \text{l.u.b.} \|Kg\| , \qquad \|g\| = 1 , \tag{2}$$

the two characterizations being equivalent. A transformation K is continuous in an $\mathscr{L}_2$-space if, whenever $\{\phi_n\}$ is a sequence in the domain of K with limit ϕ, then $K\phi_n \to K\phi$. A transformation is continuous in the entire domain of K if it is continuous at every point therein. Fortunately, a linear transformation is continuous if it is bounded and vice versa, and it is easy to show that the operator K as defined above is bounded. Indeed, by starting with the relation

$$\psi(s) = K\phi = \int K(s,t)\,\phi(t)\,dt\,,$$

we have

$$|\psi(s)|^2 = |\int K(s,t)\,\phi(t)\,dt|^2 \leqslant \int |K(s,t)|^2\,dt \int |\phi(t)|^2\,dt$$

or

$$|\psi(s)|^2 \leqslant \|\phi\|^2 \int |K(s,t)|^2\,dt\,.$$

Another integration yields

$$\|\psi\| = \|K\phi\| \leqslant \|\phi\| \left[\int\int |K(s,t)|^2\,ds\,dt\right]^{1/2},$$

which implies that

$$\|K\| \leqslant \left[\int\int |K(s,t)|^2\,ds\,dt\right]^{1/2},\qquad(3)$$

as desired.

A rather important concept in the theory of linear operators is the concept of complete continuity. An operator is described as completely continuous if it transforms any bounded set into a compact set (a set S of elements ϕ is called compact if a subsequence having a limit can be extracted from any sequence of elements of S). Obviously, a completely continuous operator is continuous (and hence bounded), but the converse is not true. Furthermore, any bounded operator K whose range is finite-dimensional is completely continuous because it transforms a bounded set in $\mathscr{L}_2(a,b)$ into a bounded finite-dimensional set which is necessarily compact. Many of the integral operators that arise in applications are completely continuous. For instance, a separable kernel $K(s,t)$,

$$K(s,t) = \sum_{i=1}^{n} a_i(s)\,b_i(t)\,,$$

where $a_i(s)$ and $b_i(t)$ are $\mathscr{L}_2$-functions, is completely continuous, as can be proved as follows. Indeed, for each $\mathscr{L}_2$-function

$$Kg = \int\left[\sum_{i=1}^{n} a_i(s)\,b_i(t)\,g(t)\right]dt = \sum_{i=1}^{n} c_i\,a_i(s)\,,$$

that is, the range of K is a finite-dimensional subspace of $\mathscr{L}_2(a,b)$. In addition,

$$\|Kg\| = \left\| \sum_{i=1}^{n} c_i a_i(s) \right\| \leqslant \sum_{i=1}^{n} |c_i| \, \|a_i\|$$

$$\leqslant \sum_{i=1}^{n} \|a_i\| \int |b_i(t)| \, |g(t)| \, dt \, . \tag{4}$$

Applying the Schwarz inequality in (4), we have

$$\|Kg\| \leqslant M \|g\| \, , \tag{5}$$

where $M = \sum_{i=1}^{n} \|a_i\| \, \|b_i\|$. This means that K is a bounded operator with finite-dimensional range and hence is completely continuous.

We can use this result to prove that an $\mathscr{L}_2$-kernel $K(s, t)$ is completely continuous. We need only the theorem that, if K can be approximated in norm by a completely continuous operator, then A is completely continuous. If we assume this theorem, then our contention is proved because an $\mathscr{L}_2$-kernel can always be approximated by separable kernels, as shown in the previous section.

Next, we prove the interesting result that the norms of K and of its adjoint K^* are equal. To this end, we appeal to the relation (7.1.12):

$$(K\phi, \psi) = (\phi, K^*\psi) \, , \tag{6}$$

which holds for each pair of $\mathscr{L}_2$-functions ϕ, ψ. Substituting $\psi = K\phi$ and applying the Schwarz inequality, we obtain

$$(K\phi, K\phi) = (\phi, K^*K\phi) \leqslant \|\phi\| \, \|K^*K\phi\| \, ,$$

where we have used the fact that $(K\phi, K\phi)$ is nonnegative real number. Hence,

$$\|K\phi\|^2 \leqslant \|K^*\| \, \|\phi\| \, \|K\phi\| \qquad \text{or} \qquad \|K\phi\| \leqslant \|K^*\| \, \|\phi\| \, .$$

This last inequality implies that $\|K\| \leqslant \|K^*\|$. The opposite inequality is obtained by setting $\phi = K^*\psi$ in (6).

In Section 7.2, we stated the property that the reciprocal of the modulus of the eigenvalue with the smallest modulus for a symmetric $\mathscr{L}_2$-kernel K is equal to the maximum value of $|(K\phi, \phi)|$ with $\|\phi\| = 1$. This property can be proved as follows. Indeed, an upper bound for the reciprocal of the eigenvalues is immediately available because, for the eigenvalue problem $\lambda K\phi = \phi$,

$$(K\phi, \phi) = (1/\lambda)(\phi, \phi) = (1/\lambda) \|\phi\|^2 \, ,$$

which implies

$$(1/\lambda)\|\phi\|^2 = (K\phi, \phi) \leqslant \|K\phi\|\,\|\phi\| \leqslant \|K\|\,\|\phi\|^2 \,. \tag{7}$$

From (3) and (7), we derive an upper bound

$$|1/\lambda| \leqslant \left[\iint K(s, t)^2 \, ds\, dt\right]^{1/2} . \tag{8}$$

When the $\mathscr{L}_2$-kernel is also symmetric, we can use the following result from the theory of operators: If K is a symmetric and completely continuous operator, at least one of the numbers $\|K\|$ or $-\|K\|$ is the reciprocal of an eigenvalue of K and no other eigenvalue of K has smaller absolute value.

By recalling the definition of $\|K\|$ and the fact that a symmetric $\mathscr{L}_2$-kernel generates a completely continuous operator, we have proved property 7 of Section 7.2. In the process, we have also proved the existence of an eigenvalue.

Suppose we have found the first eigenvalue λ_1 and corresponding eigenfunction ϕ_1 in the sequences (7.3.1) and (7.3.2). To find the next eigenvalue λ_2 and the corresponding eigenfunction ϕ_2, we shorten the kernel K by subtracting the factor $\phi_1 \phi_1{}^*/\lambda_1$ from it. Then, from Theorem 2 of Section 7.3, we find that the kernel $K^{(2)} = [K-(\phi_1 \phi_1{}^*)/\lambda_1]$ satisfies all the requirements of a symmetric $\mathscr{L}_2$-kernel. Following the above discussion, we find that at least one of the numbers $\|K_2\|$ or $-\|K_2\|$ is the reciprocal of λ_2. This process is continued until all the eigenvalues and eigenfunctions are derived. The only drawback in this process is that, to find the $(n+1)$th eigenvalue, one has to find the first n eigenvalues. This situation is remedied by the so-called maximum-minimum principle, and this is not on our agenda. The reader is referred to Courant and Hilbert [4] for this discussion.

7.9. RAYLEIGH–RITZ METHOD FOR FINDING THE FIRST EIGENVALUE

Let us take a real, nonnegative, and symmetric $\mathscr{L}_2$-kernel K. We have found that the smallest eigenvalue λ_1 is characterized by the extremal (or variational) principle

$$1/\lambda_1 = \max(K\phi, \phi), \qquad \|\phi\| = 1. \tag{1}$$

The Rayleigh–Ritz method rests on selecting a special class of trial functions of the form $\phi = \sum_{i=1}^{n} \alpha_i \psi_i(s)$, where $\{\psi_i(x)\}$ is a suitably chosen set of linearly independent functions and $\{\alpha_i\}$ are real numbers. The relation (1) implies that, to obtain a close approximation for λ_1, we must maximize the function

$$(K\phi, \phi) = \left(K\left[\sum_{i=1}^{n} \alpha_i \psi_i(s)\right], \sum_{i=1}^{n} \alpha_i \psi_i(s)\right) = \sum_{i,k}^{n} K_{ik} \alpha_i \alpha_k, \tag{2}$$

subject to

$$\left\|\sum_{i=1}^{n} (\alpha_i \psi_i)\right\| = \sum_{i=1}^{n} c_{ik} \alpha_i \alpha_k = 1, \tag{3}$$

where

$$K_{ik} = (K\psi_i, \psi_k) = (\psi_i, K\psi_k); \qquad c_{ik} = (\psi_i, \psi_k) = (\psi_k, \psi_i), \tag{4}$$

are known quantities. Thereby, we have transformed the extremal problem (1) into an extremal problem in the advanced calculus of several variables $\alpha_1, \ldots, \alpha_n$. We use the method of Lagrange multipliers and set

$$\Phi = \sum_{i,k=1}^{n} (K_{ik} \alpha_i \alpha_k - \sigma c_{ik} \alpha_i \alpha_k), \tag{5}$$

where σ is an undetermined coefficient. The extremal values of α_i are determined from the equations $\partial\Phi/\partial\alpha_i = 0$;

$$\sum_{k=1}^{n} K_{ik} \alpha_k - \sigma \sum_{i=1}^{n} c_{ik} \alpha_k = 0, \qquad i = 1, \ldots, n. \tag{6}$$

This linear and homogeneous system of equations in $\alpha_1, \ldots, \alpha_n$ will have a nontrivial solution if and only if the determinant

$$\begin{vmatrix} K_{11} - \sigma c_{11} & K_{12} - \sigma c_{12} & \cdots & K_{1n} - \sigma c_{1n} \\ \vdots & & & \\ K_{n1} - \sigma c_{n1} & K_{n2} - \sigma c_{n2} & \cdots & K_{nn} - \sigma c_{nn} \end{vmatrix} = 0. \tag{7}$$

Also note that, by multiplying (6) by α_i and summing on i, one obtains $\sigma = (K\phi, \phi)$.

The determinant (7), when expanded, yields an nth-degree polynomial in σ which can be shown to have n real, nonnegative but not necessarily distinct roots. Let σ_1 be the maximum of these roots. Then, from the above discussion, we infer that $\sigma_1 \leqslant 1/\lambda_1$. This usually gives a good approximation to λ_1. When we solve the equations (6) for the vector $(\alpha_1, \ldots, \alpha_n)$ with $\sigma = \sigma_1$ and evaluate $\phi = \sum_{i=1}^{n} \alpha_i \psi_i$, it emerges that ϕ is usually not a good approximation for ϕ_1.

For the particular case when the trial functions $\psi_i(x)$ are orthonormal in the given interval, the computation is considerably simplified because then $c_{ij} = \delta_{ij}$. The relations (6) and (7) take the simple forms

$$\sum_{k=1}^{n} K_{ik} \alpha_k - \sigma \alpha_i = 0, \qquad i = 1, 2, \ldots, n \tag{8}$$

and

$$\begin{vmatrix} K_{11}-\sigma & K_{12} & \cdots & K_{1n} \\ \vdots & & & \\ K_{n1} & K_{n2} & & K_{nn}-\sigma \end{vmatrix} = 0, \tag{9}$$

respectively. We illustrate this method by two examples.

Example 1.   Find the first eigenvalue of the integral equation

$$g(s) - \lambda \int_0^1 K(s,t) g(t) \, dt = 0, \tag{10}$$

where

$$K(s,t) = \begin{cases} \tfrac{1}{2}s(2-t), & s < t, \\ \tfrac{1}{2}t(2-s), & s > t, \end{cases} \tag{11}$$

which can be shown to be a positive kernel. This integral equation can be proved to be equivalent to a simple ordinary differential equation (see Example 2 in Section 5.3. The exact value of the smallest eigenvalue is 4.115.)

To apply the Rayleigh–Ritz procedure, we take two trial functions:

$$\psi_n(s) = \sqrt{2} \sin n\pi s, \qquad n = 1, 2, \tag{12}$$

which are orthonormal in the interval $(0, 1)$. Proceeding as above, we have

$$K_{11} = 2/\pi^2, \qquad K_{12} = K_{21} = -1/2\pi^2, \qquad K_{22} = 1/2\pi^2. \tag{13}$$

The relation (9) gives for this special case

$$\begin{vmatrix} (2/\pi^2) - \sigma & -1/2\pi^2 \\ -1/2\pi^2 & (1/2\pi^2) - \sigma \end{vmatrix} = 0 . \tag{14}$$

The largest root of (14) is $\sigma = 2.15/\pi^2$. Thus, $\lambda_1 \sim \pi^2/2.15 = 4.59$. By including more functions in the sequence (12), we can improve the approximation progressively.

Example 2. Find an approximation for the smallest eigenvalue of the positive-definite symmetric kernel

$$K(s,t) = \begin{cases} s, & s < t \\ t, & s > t \end{cases} \tag{15}$$

in the basic interval $(0, 1)$.

To use the Rayleigh–Ritz method, we take two trial functions

$$\psi_1(s) = 1 , \qquad \psi_2(s) = (2s-1) , \tag{16}$$

which are orthogonal but not orthonormal [they are Legendre polynomials $P_0(2s-1)$, $P_1(2s-1)$]. Thus,

$$c_{11} = 1 , \qquad c_{12} = c_{21} = 0 , \qquad c_{22} = 1/3 ,$$
$$K_{11} = 1/3 , \qquad K_{12} = K_{21} = 1/12 , \qquad K_{22} = 1/30 . \tag{17}$$

Substituting (17) in (7) and evaluating the determinant, we have $\sigma^2 - (13/30)\sigma + (1/80) = 0$. The largest root is $\sigma_1 = (1/60)[13 + (124)^{1/2}]$. Thus, the smallest eigenvalue is $\lambda_1 = 1/\sigma_1 \sim 2.4859$, which compares favorably with the exact value 2.4674.

EXERCISES

1. Show that, if $(K\phi, \phi)$ is real for all ϕ, then K is a symmetric Fredholm operator.

2. Determine the iterated kernels for the symmetric kernel

$$K(s,t) = \sum_{k=1}^{\infty} k^{-1} \sin k\pi s \sin k\pi t .$$

3. Show that the kernel $K(s, t)$, $0 \leqslant s, t \leqslant 1$,

$$K(s,t) = \begin{cases} s(1-t), & s < t, \\ t(1-s), & s > t, \end{cases}$$

has the bilinear form

$$K(s,t) = 2 \sum_{k=1}^{\infty} \frac{\sin k\pi s \sin k\pi t}{(k\pi)^2}.$$

4. Use the result in Exercise 3 and show that

$$\sum_{n=1}^{\infty} \frac{1}{n^2} = \frac{\pi^2}{6}.$$

5. Consider the eigenvalue problem

$$g(s) = \lambda \int_{-1}^{1} (1 - |s-t|) g(t) \, dt.$$

Differentiate under the integral sign to obtain the corresponding differential equation and boundary conditions. Show that the kernel of this integral equation is positive.

6. Determine the eigenvalues and eigenfunctions of the symmetric kernel $K(s, t) = \min(s, t)$ in the basic interval $0 \leqslant s, t \leqslant 1$.

7. Use the Hilbert–Schmidt theorem to solve the symmetric integral equations as given in Examples 1–3 and 6 in Section 2.2.

8. Apply the Gram–Schmidt process to orthogonalize $1, s, s^2, s^3$ in the interval $-1 \leqslant s \leqslant 1$. Use this result to find the eigenvalues and eigenfunctions of the symmetric kernel $K(s, t) = 1 + st + s^2 t^2 + s^3 t^3$.

9. Consider the kernel $K(s, t) = \log[1 - \cos(s - t)]$, $0 \leqslant s, t \leqslant 2\pi$. Show that (a) it is a symmetric $\mathcal{L}_2$-kernel; (b) the following holds:

$$K(s,t) = -(\log 2) + 2 \log \left[1 - e^{i(s-t)}\right]$$

$$= -(\log 2) - 2 \sum_{n=1}^{\infty} \frac{\cos ns \cos nt}{n} - 2 \sum_{n=1}^{\infty} \frac{\sin ns \sin nt}{n};$$

and (c) its eigenvalues are $\lambda_0 = -1/(2\pi \log 2)$, $\lambda_n = -n/2\pi$, $n = 1, 2, \ldots$, with eigenfunctions $\phi_0(s) = C$, $\phi_n(s) = A \cos ns + B \sin ns$, where A, B, and C are constants.

10. By combining Sections 5.5 and 7.4, show that the eigenfunctions of any self-adjoint differential system of the second order form a complete set.

11. Prove that, for a square-integrable function, the Fourier transform preserves norms.

SINGULAR INTEGRAL EQUATIONS

8.1. THE ABEL INTEGRAL EQUATION

An integral equation is called singular if either the range of integration is infinite or the kernel has singularities within the range of integration. Such equations occur rather frequently in mathematical physics and possess very unusual properties. For instance, one of the simplest singular integral equations is the Abel integral equation

$$f(s) = \int_0^s [g(t)/(s-t)^\alpha] \, dt, \qquad 0 < \alpha < 1, \tag{1}$$

which arises in the following problem in mechanics. A material point moving under the influence of gravity along a smooth curve in a vertical plane takes the time $f(s)$ to move from the vertical height s to a fixed point 0 on the curve. The problem is to find the equation of that curve. Equation (1) with $\alpha = 1/2$ is the integral-equation formulation of this problem.

The integral equation (1) is readily solved by multiplying both sides by the factor $ds/(u-s)^{1-\alpha}$ and integrating it with respect to s from 0 to u:

$$\int_0^u \frac{f(s)\,ds}{(u-s)^{1-\alpha}} = \int_0^u \frac{ds}{(u-s)^{1-\alpha}} \int_0^s \frac{g(t)\,dt}{(s-t)^\alpha}. \tag{2}$$

The double integration on the right side of the above equation is so written that first it is to be integrated in the t direction from 0 to s and then the resulting single integral is to be integrated in the s direction from 0 to u. The region of integration therefore is the triangle lying below the diagonal $s = t$. We change the order of integration so that we first integrate from $s = t$ to $s = u$ and afterwards in the t direction from $t = 0$ to $t = u$. Equation (2) then becomes

$$\int_0^u \frac{f(s)\,ds}{(u-s)^{1-\alpha}} = \int_0^u g(t)\,dt \int_t^u \frac{ds}{(u-s)^{1-\alpha}(s-t)^\alpha}. \tag{3}$$

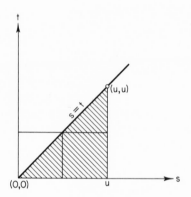

Figure 8.1

To evaluate the integral

$$\int_t^u \frac{ds}{(u-s)^{1-\alpha}(s-t)^\alpha},$$

one sets $y = (u-s)/(u-t)$, and obtains

$$\int_t^u (u-s)^{\alpha-1}(s-t)^{-\alpha}\, ds = \int_0^1 y^{\alpha-1}(1-y)^{-\alpha}\, dy = \pi/\sin \alpha\pi ,$$

where we have used the value of the Eulerian beta function $B(\alpha, 1-\alpha)$ $= \pi/\sin \alpha\pi$. Substituting this result in (3), we have

$$\frac{\sin \alpha\pi}{\pi}\int_0^u \frac{f(s)\, ds}{(u-s)^{1-\alpha}} = \int_0^u g(t)\, dt ,$$

which, when differentiated with respect to u, and then changing u to t, gives the required solution:

$$g(t) = \frac{\sin \alpha\pi}{\pi}\frac{d}{dt}\left[\int_0^t f(s)(t-s)^{\alpha-1}\, ds\right] . \tag{4}$$

The integral equation (1) is a special case of the singular integral equation [18]

$$f(s) = \int_a^s \frac{g(t)\, dt}{[h(s)-h(t)]^\alpha} , \quad 0 < \alpha < 1 , \tag{5}$$

where $h(t)$ is a strictly monotonically increasing and differentiable function in (a, b), and $h'(t) \neq 0$ in this interval. To solve this, we consider the integral

$$\int_a^s \frac{h'(u)f(u)\, du}{[h(s)-h(u)]^{1-\alpha}} ,$$

and substitute for $f(u)$ from (5). This gives

$$\int_a^s \int_a^u \frac{g(t)\, h'(u)\, dt\, du}{[h(u)-h(t)]^\alpha\, [h(s)-h(u)]^{1-\alpha}} ,$$

which, by change of the order of integration, becomes

$$\int_a^s g(t)\, dt \int_t^s \frac{h'(u)\, du}{[h(u) - h(t)]^\alpha [h(s) - h(u)]^{1-\alpha}} .$$

The inner integral is easily proved to be equal to the beta function $B(\alpha, 1-\alpha)$. We have thus proved that

$$\int_a^s \frac{h'(u)f(u)\, du}{[h(s) - h(u)]^{1-\alpha}} = \frac{\pi}{\sin \alpha\pi} \int_a^s g(t)\, dt , \qquad (6)$$

and by differentiating both sides of (6), we obtain the solution

$$g(t) = \frac{\sin \alpha\pi}{\pi} \frac{d}{dt} \int_a^t \frac{h'(u)f(u)\, du}{[h(t) - h(u)]^{1-\alpha}} . \qquad (7)$$

Similarly, the integral equation

$$f(s) = \int_s^b \frac{g(t)\, dt}{[h(t) - h(s)]^\alpha} , \qquad 0 < \alpha < 1 , \qquad (8)$$

and $a < s < b$, with $h(t)$ a monotonically increasing function, has the solution

$$g(t) = -\frac{\sin \alpha\pi}{\pi} \frac{d}{dt} \int_t^b \frac{h'(u)f(u)\, du}{[h(u) - h(t)]^{1-\alpha}} . \qquad (9)$$

We close this section with the remark that a Fredholm integral equation with a kernel of the type

$$K(s, t) = H(s, t)/|t-s|^\alpha , \qquad 0 < \alpha < 1 , \qquad (10)$$

where $H(s, t)$ is a bounded function, can be transformed to a kernel which is bounded. It is done by the method of iterated kernels. Indeed, it can be shown [11, 15, 20] that, if the singular kernel has the form as given by the relation (10), then there always exists a positive integer p_0, dependent on α, such that, for $p > p_0$, the iterated kernel $K_p(s, t)$ is bounded. For this reason, the kernel (10) is called weakly singular.

Note that, for this hypothesis, the condition $\alpha < 1$ is essential. For the important case $\alpha = 1$, the integral equation differs radically from the equations considered in this section. Moreover, we need the notion of Cauchy principal value for this case. But, before considering the case $\alpha = 1$, let us give some examples for the case $\alpha < 1$.

8.2. EXAMPLES

Example 1. Solve the integral equation

$$s = \int_0^s \frac{g(t)\, dt}{(s-t)^{1/2}}.$$ (1)

Comparing this with integral equation (8.1.1), we find that $f(s) = s$, $\alpha = 1/2$. Substituting these values in (8.1.4), there results the solution:

$$g(t) = \frac{1}{\pi} \frac{d}{dt} \left[\int_0^t \frac{s}{(t-s)^{1/2}}\, ds \right]$$

$$= \frac{1}{\pi} \frac{d}{dt} \left[-\frac{2}{3}(s+2t)(t-s)^{1/2} \right]_0^t$$

$$= \frac{1}{\pi} \frac{d}{dt} \left[\frac{4}{3} t^{3/2} \right] = \frac{2t^{1/2}}{\pi}.$$ (2)

Example 2. Solve the integral equation

$$f(s) = \int_a^s \frac{g(t)\, dt}{(\cos t - \cos s)^{1/2}}, \qquad 0 \leqslant a < s < b \leqslant \pi.$$ (3)

Comparing (8.1.5) and (3), we see that $\alpha = 1/2$, and $h(t) = 1 - \cos t$, a strictly monotonically increasing function in $(0, \pi)$. Substituting this value for $h(u)$ in (8.1.7), we have the required solution

$$g(t) = \frac{1}{\pi} \frac{d}{dt} \left[\int_a^t \frac{(\sin u) f(u)\, du}{(\cos u - \cos t)^{1/2}} \right], \qquad a < t < b.$$ (4)

Similarly, the integral equation

$$f(s) = \int_s^b \frac{g(t)\, dt}{(\cos s - \cos t)^{\frac{1}{2}}}, \qquad 0 \leqslant a < s < b \leqslant \pi, \qquad (5)$$

has the solution

$$g(t) = -\frac{1}{\pi}\frac{d}{dt}\left[\int_t^b \frac{(\sin u)f(u)\, du}{(\cos t - \cos u)^{\frac{1}{2}}}\right], \qquad a < t < b. \qquad (6)$$

Example 3.   Solve the integral equations

$$(a) \qquad f(s) = \int_a^s \frac{g(t)\, dt}{(s^2 - t^2)^\alpha}, \qquad 0 < \alpha < 1 ; \quad a < s < b, \qquad (7)$$

and

$$(b) \qquad f(s) = \int_s^b \frac{g(t)\, dt}{(t^2 - s^2)^\alpha}, \qquad 0 < \alpha < 1 ; \quad a < s < b. \qquad (8)$$

From (8.1.5) and (7), we find that $h(t) = t^2$, which is a strictly monotonic function. The solution, therefore, follows from (8.1.7):

$$g(t) = \frac{2\sin\alpha\pi}{\pi}\frac{d}{dt}\int_a^t \frac{uf(u)\, du}{(t^2 - u^2)^{1-\alpha}}, \qquad a < t < b. \qquad (9)$$

Similarly, the solution of the integral equation (8) is

$$g(t) = -\frac{2\sin\alpha\pi}{\pi}\frac{d}{dt}\int_t^b \frac{uf(u)\, du}{(u^2 - t^2)^{1-\alpha}}, \qquad a < t < b. \qquad (10)$$

The results (9) and (10) remain valid when a tends to 0 and b tends to $+\infty$. Hence, the solution of the integral equation

$$f(s) = \int_0^s \frac{g(t)\, dt}{(s^2 - t^2)^\alpha}, \qquad 0 < \alpha < 1, \qquad (11)$$

is

$$g(t) = \frac{2\sin\alpha\pi}{\pi} \frac{d}{dt} \int_0^t \frac{uf(u)\,du}{(t^2-u^2)^{1-\alpha}}. \tag{12}$$

Similarly, the solution of the integral equation

$$f(s) = \int_s^\infty \frac{g(t)\,dt}{(t^2-s^2)^\alpha}, \qquad 0 < \alpha < 1, \tag{13}$$

is

$$g(t) = -\frac{2\sin\alpha\pi}{\pi} \frac{d}{dt} \int_t^\infty \frac{uf(u)\,du}{(u^2-t^2)^{1-\alpha}}. \tag{14}$$

8.3. CAUCHY PRINCIPAL VALUE FOR INTEGRALS

The theory of Riemann integration is based on the assumption that the range of integration is finite and that the integrand is bounded. For the integration of an unbounded function or an infinite range of integration, the concept of improper integrals is introduced.

Consider a function $f(s)$, defined in the interval $a \leqslant s \leqslant b$, which is unbounded in the neighborhood of a point c, $a < c < b$, but is integrable in each of the intervals $(a, c-\varepsilon)$ and $(c+\eta, b)$ where ε and η are arbitrary, small positive numbers. Then, the limit

$$\int_a^b f(s)\,ds = \lim_{\substack{\varepsilon\to 0 \\ \eta\to 0}} \left[\int_a^{c-\varepsilon} f(s)\,ds + \int_{c+\eta}^b f(s)\,ds \right], \tag{1}$$

if it exists, is called the improper integral of the function $f(s)$ in the range (a, b). Here, it is implied that ε and η tend to zero independently. But it may happen that the limit (1) does not exist when ε and η tend to zero independently of each other, but it exists if ε and η are related. The classic example is the function $f(s) = 1/(s-c)$, $a < c < b$; the limit (1) in this case is

$$\int_a^{c-\varepsilon} \frac{ds}{s-c} + \int_{c+\eta}^b \frac{ds}{s-c} = \log\frac{b-c}{c-a} + \log\frac{\varepsilon}{\eta} \,.$$

If ε and η tend to zero independently of each other, then the quantity $\log(\varepsilon/\eta)$ will vary arbitrarily. However, if ε and η are related, then the above limit exists. In the special case $\varepsilon = \eta$, this limit is

$$\int_a^b \frac{ds}{s-c} = \log\frac{b-c}{c-a} \tag{2}$$

and is called the Cauchy principal value or Cauchy principal integral.

The same definition applies to a general function $f(s)$. The Cauchy principal value of a function $f(s)$ that becomes infinite at an interior point $x = c$ of the range of integration (a, b) is the limit

$$\lim_{\varepsilon \to 0} \left(\int_a^{c-\varepsilon} + \int_{c+\varepsilon}^b \right) f(s) \, ds \,, \tag{3}$$

where

$$0 < \varepsilon \leqslant \min(c-a, b-c) \,.$$

Such a limit is usually denoted as $P\int_a^b f(s)\, ds$ or $\int_{*a}^{*b} f(s)\, ds$. We shall use the latter symbol in the sequel.

A similar definition for the Cauchy principal value is given for integrals with an infinite range of integration. For instance, the limit

$$\int_{-\infty}^\infty f(s) \, ds = \lim_{\substack{A \to \infty \\ B \to \infty}} \int_{-A}^B f(s) \, ds$$

may not exist when A and B tend to infinity independently of each other, but the limit exists when $A = B$. This limit,

$$\lim_{A \to \infty} \int_{-A}^A f(s) \, ds \,, \tag{4}$$

is called the Cauchy principal value. The limits (3) and (4) are also called singular integrals.

If a function $f(s)$ satisfies certain regularity conditions, then the above-mentioned singular integrals exist. One such concept of regularity is the Hölder condition. A function $f(s)$ is said to satisfy the Hölder condition if there exist constants k and α, $0 < \alpha \leqslant 1$, such that, for every pair of points s_1, s_2 lying in the range $a \leqslant s \leqslant b$, we have

$$|f(s_1) - f(s_2)| < k |s_1 - s_2|^\alpha . \tag{5}$$

Such a function is also said to be Hölder continuous. The special case $\alpha = 1$ is often called the Lipschitz condition.

It is not hard to prove that, if $f(s)$ is Hölder continuous, then the singular integral

$$\int_a^{*b} [f(t)/(t-s)] \, dt \tag{6}$$

exists. Indeed, the relation (6) can be split as

$$f(s) \int_a^b \frac{dt}{t-s} + \int_a^b \frac{f(t) - f(s)}{t-s} \, dt . \tag{7}$$

The first integral has the principal value as proved by the relation (2). In the second integral, the integrand is such that

$$\left| \frac{f(t) - f(s)}{t-s} \right| < k |t-s|^{\alpha - 1} .$$

Therefore, this integral exists as an improper integral for $\alpha < 1$ and as a Riemann integral for $\alpha = 1$.

The function $f_1(s)$ defined by the singular integral (6),

$$f_1(s) = \int_a^{*b} [f(t)/(t-s)] \, dt ,$$

has the following property, which we state without proof. If $f(s)$ is Hölder continuous with exponent α, $\alpha < 1$, then $f_1(s)$ is also Hölder continuous in every closed interval (a_1, b_1), where $a < a_1 \leqslant x \leqslant b_1 < b$. When $f(s)$ is Hölder continuous with $\alpha = 1$, then $f_1(s)$ is Hölder continuous with exponent β, which is an arbitrary positive number less than unity.

The Hölder condition can be extended to functions of more than one variable. For example, the kernel $K(s, t)$ is Hölder continuous with respect to both variables if there exist constants k and α, $0 < \alpha \leqslant 1$, such that

$$|K(s_1, t_1) - K(s_2, t_2)| < k\left[|s_1 - s_2|^\alpha + |t_1 - t_2|^\alpha\right], \tag{8}$$

where (s_1, t_1) and (s_2, t_2) lie within the range of definition.

The Cauchy principal value for contour integrals is also defined in a similar fashion. A contour integral of a complex-valued function with a pole c on the contour strictly does not exist. However, it may have the Cauchy principal value if this concept is extended for this case. For this purpose, let L be a closed or open regular curve (i.e., it has continuous curvature at every point) (see Figure 8.2). Enclose the point c by a

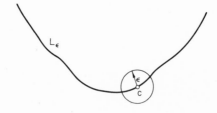

Figure 8.2

small circle of radius ε with center at c. Let L_ε denote the part of the contour outside this circle. If a complex-valued function $f(z)$ is integrable along L_ε, however small the positive number ε, then the limit

$$\lim_{\varepsilon \to 0} \int_{L_\varepsilon} f(z)\, dz \, ,$$

if it exists, is called the Cauchy principal value and is denoted as

$$\int_L^* f(z)\, dz$$

or

$$P \int_L f(z)\, dz.$$

We shall be interested in the contour integrals of the Cauchy type,

$$\int_L [f(\tau)/(\tau-z)]\, d\tau \,, \tag{9}$$

in the sequel. It is known in the theory of functions of a complex variable that, if $f(z)$ satisfies the Hölder condition

$$|f(z_1)-f(z_2)| < k\,|z_1-z_2|^\alpha \,, \tag{10}$$

where z_1, z_2 is any pair of points on L, while k and α are constants such that $0 < \alpha \leqslant 1$, then the integral (9) exists for all points z on the curve L, except perhaps its end points. The function $f_1(z)$ defined by the integral (9) is also Hölder continuous, with similar properties as given for the case of the corresponding real functions.

The definition (10) can be extended to complex-valued functions of more than one variable as was done for the real-valued functions above. Incidentally, the function $f(\tau)$ occurring in the integral (9) is called the density of the Cauchy integral.

8.4. THE CAUCHY-TYPE INTEGRALS

The integral equation

$$f(z) = \frac{1}{2\pi i}\int_L \frac{g(\tau)}{\tau-z}\, d\tau \,, \tag{1}$$

where L is a regular curve, is called a Cauchy-type integral. We shall first study the case when L is a closed contour. For the discussion of the integral equation (1), we need a result from the theory of complex-valued functions, which we state without proof.

Let $g(\tau)$ be a Hölder-continuous function of a point on a regular closed contour L and let a point z tend, in an arbitrary manner, from inside or outside the contour L, to the point t on this contour; then the integral (1) tends to the limit [3, 15]

$$f^+(t) = \frac{1}{2}g(t) + \frac{1}{2\pi i}\int_L^* \frac{g(\tau)}{\tau-t}\, d\tau \tag{2}$$

or

$$f^-(t) = -\frac{1}{2}g(t) + \frac{1}{2\pi i}\int_L^* \frac{g(\tau)}{\tau - t}\,d\tau \,, \tag{3}$$

respectively. The formulas (2) and (3) are known as Plemelj formulas. It is interesting to compare them with the formulas (6.2.10) and (6.2.11). Incidentally, we follow the standard convention of counterclockwise traversal of the closed contour L. This means that the first boundary value $f^+(t)$ relates to the value of the Cauchy integral inside the region bounded by L, while the second boundary value $f^-(t)$ relates to the value in the outside region.

Let

$$g_1(t) = \frac{1}{2\pi i}\int_L^* \frac{g(\tau)}{\tau - t}\,d\tau \,, \tag{4}$$

$$g_2(t) = \frac{1}{2\pi i}\int_L^* \frac{g_1(\tau)}{\tau - t}\,d\tau \tag{5}$$

be two singular integrals, where g and g_1 are Hölder-continuous functions and L is a closed contour. Can we compound these two integrals and obtain an iterated integral connecting the functions g_2 and g? The answer is in the affirmative. To prove this assertion, consider the Cauchy-type integrals

$$f(z) = \frac{1}{2\pi i}\int_L \frac{g(\tau)}{\tau - z}\,d\tau \tag{6}$$

and

$$f_1(z) = \frac{1}{2\pi i}\int_L \frac{g_1(\tau)}{\tau - z}\,d\tau \,. \tag{7}$$

Using the Plemelj formula (2) and the integrals (6) and (7), we obtain the limiting values

$$f^+(t) = \frac{1}{2}g(t) + \frac{1}{2\pi i}\int_L^* \frac{g(\tau)}{\tau - t}\,d\tau \,, \tag{8}$$

& Kanwal's proof too long: $g = \frac{1}{2\pi i}\int \frac{g(t)}{t-t}\,d\tau \,, \quad g_2 = \frac{1}{2\pi i}\,P\int \frac{g_1(t)}{t-t}\,d\tau$

$f(t) = \frac{1}{2\pi i}\int \frac{g(t)\,dt}{t-t}$

$f_1(t) = \frac{1}{2\pi i}\int \frac{g_1(t)\,dt}{t-t}$ $f \to f^+ = \frac{1}{2}g + \frac{1}{2\pi i}P\int \frac{g}{t-t}\,d\tau$ as $t \to L$ *from inside.*

$\quad\quad\quad\quad\quad n\quad g_1 = f^+ - \frac{1}{2}g$

$\quad\quad\quad\quad\quad g_2 = f_1^+ - \frac{1}{2}g_1 = f_1^+ - \frac{1}{2}f^+ + \frac{1}{4}g$

But $f_1 = \frac{1}{2\pi i}\int \frac{(f^+ - \frac{1}{2}g)}{(t-t)}\,d\tau = f - \frac{1}{2}f = \frac{1}{2}f$ *or* $f_1^+ = \frac{1}{2}f^+$ *so* $g_2 = \frac{1}{4}g.$

[*If g is Hölder continuous, f is regular in L*].

$$f_1{}^+(t) = \frac{1}{2}\, g_1(t) + \frac{1}{2\pi i} \int\limits_L{}^{*} \frac{g_1(\tau)}{\tau - t}\, d\tau \ . \tag{9}$$

Comparing the relations (4) and (8) on one hand and the relations (5) and (9) on the other hand, we obtain

$$g_1(t) = f^+(t) - \tfrac{1}{2}g(t) , \tag{10}$$

$$g_2(t) = f_1{}^+(t) - \tfrac{1}{2}g_1(t) . \tag{11}$$

Substituting (10) in (7) yields

$$f_1(z) = \frac{1}{2\pi i} \int\limits_L \frac{f^+(\tau)}{\tau - z}\, d\tau - \frac{1}{4\pi i} \int\limits_L \frac{g(\tau)}{\tau - z}\, d\tau \ . \tag{12}$$

The value of the first integral in equation (12) is $f(z)$ because its density $f^+(\tau)$ is the limiting value of the function $f(z)$, which is regular inside L, and therefore we can use the Cauchy integral formula. The second integral is one-half of the integral in (6). Hence, $f_1(z) = \tfrac{1}{2}f(z)$, which implies that

$$f_1{}^+(t) = \tfrac{1}{2}f^+(t) . \tag{13}$$

From (10), (11), and (13), it follows that

$$g_2(t) = \tfrac{1}{2}f^+(t) - \tfrac{1}{2}[f^+(t) - \tfrac{1}{2}g(t)] = \tfrac{1}{4}g(t) . \tag{14}$$

Finally, from (4), (5), and (14), we have the required iterated integral equation: ✒

$$\frac{1}{(2\pi i)^2} \int\limits_L{}^{*} \frac{d\tau_1}{\tau_1 - t} \int\limits_L{}^{*} \frac{g(\tau)}{\tau - \tau_1}\, d\tau = \frac{1}{4}g(t) , \tag{15}$$

the so-called Poincaré–Bertrand transformation formula.

It is interesting to note that, in the formula (15), it is not permissible to change the order of integration. Indeed, if we change the order of integration, then the left side of (15) gives

$$\frac{1}{(2\pi i)^2} \int\limits_L{}^{*} g(\tau)\, d\tau \int\limits_L{}^{*} \frac{d\tau_1}{(\tau - \tau_1)(\tau_1 - t)} \ . \tag{16}$$

But

$$\int\limits_{L}^{*} \frac{d\tau_1}{(\tau - \tau_1)(\tau_1 - t)} = \frac{1}{\tau - t}\left[\int\limits_{L}^{*} \frac{d\tau_1}{\tau_1 - t} - \int\limits_{L}^{*} \frac{d\tau_1}{\tau_1 - \tau}\right] = 0 \, ,$$

where we have used the Plemelj formula (2), which gives for the present case

$$\int\limits_{L}^{*} \frac{d\tau_1}{\tau_1 - t} = \int\limits_{L}^{*} \frac{d\tau_1}{\tau_1 - \tau} = \pi i \, .$$

Thus, the relation (16) is equal to zero and not $\frac{1}{4}g(t)$ as in (15).

8.5. SOLUTION OF THE CAUCHY-TYPE SINGULAR INTEGRAL EQUATION

(i) CLOSED CONTOUR. The problem is to solve the integral equation of the second kind

$$ag(t) = f(t) - \frac{b}{\pi i}\int\limits_{L}^{*} \frac{g(\tau)}{\tau - t} d\tau \, , \tag{1}$$

where a and b are given complex constants, $g(\tau)$ is a Hölder-continuous function, and L is a regular contour. A fortunate aspect of this integral equation is that it can be solved simultaneously for the cases $a \neq 0$, and $a = 0$, so that the solution of the integral equation of the first kind follows as a limiting case.

To solve (1), we write it in the operator form

$$Lg = ag(t) + \frac{b}{\pi i}\int\limits_{L}^{*} \frac{g(\tau)}{\tau - t} d\tau = f(t) \, , \tag{2}$$

and define an "adjoint" operator

$$M\phi = a\phi(t) - \frac{b}{\pi i} \int\limits_{L}^{*} \frac{\phi(\tau)}{\tau - t} \, d\tau \,. \tag{3}$$

From (2) and (3), it follows that

$$MLg = a\left[ag(t) + \frac{b}{\pi i} \int\limits_{L}^{*} \frac{g(t)}{\tau - t} \, d\tau \right]$$

$$- \frac{b}{\pi i} \int\limits_{L}^{*} \frac{d\tau_1}{\tau_1 - t} \left[ag(\tau_1) + \frac{b}{\pi i} \int\limits_{L}^{*} \frac{g(\tau) \, d\tau}{\tau - \tau_1} \right] = Mf \,. \tag{4}$$

Using the Poincaré–Bertrand formula (8.5.15) and after a slight simplification, equation (4) becomes

$$g(t) = \frac{a}{a^2 - b^2} f(t) - \frac{b}{(a^2 - b^2) \pi i} \int\limits_{L}^{*} \frac{f(\tau)}{\tau - t} \, d\tau \,, \tag{5}$$

where it is assumed that $a^2 - b^2 \neq 0$. Substituting (5) back in the integral equation (1), it is found that the function $g(t)$ indeed satisfies the original integral equation.

The solution of the Cauchy-type integral equation of the first kind,

$$f(t) = \frac{b}{\pi i} \int\limits_{L}^{*} \frac{g(t)}{\tau - t} \, d\tau, \tag{6}$$

follows by setting $a = 0$ in (5):

$$g(t) = \frac{1}{b\pi i} \int\limits_{L}^{*} \frac{f(\tau)}{\tau - t} \, d\tau \,. \tag{7}$$

For $b = 1$, equations (6) and (7) take the form

$$\frac{1}{\pi i} \int\limits_{L}^{*} \frac{g(\tau) \, d\tau}{\tau - t} = f(t) \,, \qquad \frac{1}{\pi i} \int\limits_{L}^{*} \frac{f(\tau) \, d\tau}{\tau - t} = g(t) \,, \tag{8}$$

which displays the reciprocity of these relations.

(ii) UNCLOSED CONTOURS AND THE RIEMANN–HILBERT PROBLEM. The analysis of the case (i) is based on the application of the Poincaré–Bertrand formula. When the contour L is not closed, then this formula is not applicable and new methods have to be devised to solve the integral equation (1). However, the Plemelj formulas (8.4.2) and (8.4.3) are valid for an arc also when we define the plus and minus directions as follows. Supplement the arc L with another arc L' so as to form a closed contour $L+L'$. Then, the interior and exterior of this closed contour stand for the plus and minus directions. Thus, we have

$$f^+(t) = \frac{1}{2} g(t) + \frac{1}{2\pi i} \int_L^* \frac{g(\tau)}{\tau - t} d\tau , \tag{9}$$

$$f^-(t) = -\frac{1}{2} g(t) + \frac{1}{2\pi i} \int_L^* \frac{g(\tau)}{\tau - t} d\tau , \tag{10}$$

for an arc L. These formulas can also be written as

$$g(t) = f^+(t) - f^-(t) , \tag{11}$$

$$\frac{1}{\pi i} \int_L^* \frac{g(\tau)}{\tau - t} dt = f^+(t) + f^-(t) . \tag{12}$$

Now, suppose that a function $w(t)$ is prescribed on an arc L and that it satisfies the Hölder condition on L. It is required to find a function $W(z)$ analytic for all points z on L such that it satisfies the boundary (or jump) condition

$$W^+(t) - W^-(t) = w(t) , \qquad t \in L . \tag{13}$$

The formula (11) obviously helps us in evaluating such a function $W(z)$. The problem posed in (13) is a special case of the so-called Riemann–Hilbert problem, which requires the determination of a function $W(z)$ analytic for all points z not lying on L such that, for t on L,

$$W^+(t) - Z(t)W^-(t) = w(t) , \tag{14}$$

where $w(t)$ and $Z(t)$ are given complex-valued functions.

It follows by substituting the formulas (11) and (12) in the integral equation

$$ag(t) = F(t) - \frac{b}{\pi i} \int_L^* \frac{g(\tau)}{\tau - t} d\tau , \qquad (15)$$

that the solution of this integral equation is reduced to solving the Riemann–Hilbert problem

$$(a+b)f^+(t) - (a-b)f^-(t) = F(t) . \qquad (16)$$

We shall content ourselves with merely writing down the solution of (1). For details on the Riemann–Hilbert problem the reader is referred elsewhere [3, 13, 15].

Let L be a regular unclosed curve; then the solution of the singular integral equation (1) is

$$g(t) = \frac{a}{a^2 - b^2} f(t) - \frac{b}{(a^2 - b^2)\pi i} \left(\frac{t-\alpha}{t-\beta}\right)^m$$

$$\times \int_L^* \left(\frac{\tau-\beta}{\tau-\alpha}\right)^m f(\tau) \frac{d\tau}{\tau - t} + \frac{c}{(t-\alpha)^{1-m}(t-\beta)^m} , \qquad (17)$$

where α and β are the beginning and end points of the contour L and the number m is defined as

$$m = \frac{1}{2\pi i} \log \frac{a+b}{a-b} .$$

The quantity c is an arbitrary constant and is suitably chosen so that $g(t)$ is bounded at α or at β.

In particular, the solution of the integral equation of the first kind (we can put $b = 1$ without any loss of generality),

$$f(t) = \frac{1}{i\pi} \int_L^* \frac{g(\tau)}{\tau - t} d\tau , \qquad (18)$$

is obtained from (17) by setting $a = 0$, $b = 1$. Then, $m = 1/2$ and

$$g(t) = \frac{1}{\pi i}\left(\frac{t-\alpha}{t-\beta}\right)^{\frac{1}{2}} \int\limits_{L}^{*}\left(\frac{\tau-\beta}{\tau-\alpha}\right)^{\frac{1}{2}} \frac{f(\tau)}{\tau-t}\, dt + \frac{c}{[(t-\alpha)(t-\beta)]^{\frac{1}{2}}}\, . \qquad (19)$$

8.6. THE HILBERT KERNEL

A kernel of the form

$$K(s, t) = \cot[(t-s)/2]\, ,$$

where s and t are real variables, is called the Hilbert kernel and is closely connected with the Cauchy kernel. In fact, the integral equation

$$g(s) = f(s) - \lambda \int\limits_{0}^{*2\pi} F(s, t)\,\{\cot[(t-s)/2]\}\, g(t)\, dt\, , \qquad (1)$$

where $f(s)$ and $F(s, t)$ are given continuous functions of period 2π, is equivalent to the Cauchy-type integral equation

$$g(\zeta) = f(\zeta) - \lambda \int\limits_{L}^{*} [G(\zeta, \tau)/(\tau-\zeta)]\, g(\tau)\, d\tau\, , \qquad (2)$$

where ζ and τ are complex variables and the contour L is the circumference of the unit disk with center at the point $z = 0$.

Let ζ and τ denote the points of the boundary L corresponding to the arguments s and t, respectively:

$$\zeta = e^{is}\, , \qquad \tau = e^{it}\, ,$$

so that

$$\frac{d\tau}{\tau-\zeta} = i\,\frac{e^{it}\, dt}{e^{it}-e^{is}} = \left[\frac{1}{2}\cot\left(\frac{t-s}{2}\right) + \frac{i}{2}\right] dt\, . \qquad (3)$$

Therefore,

$$\cot\left(\frac{t-s}{2}\right) dt = \frac{2\, d\tau}{\tau-\zeta} - \frac{d\tau}{\tau} = \frac{\tau+\zeta}{\tau-\zeta}\frac{d\tau}{\tau}\, , \qquad (4)$$

and equation (1) takes the form (2).

The Hilbert kernel (1) is also related to the Poisson kernel in the integral representation formula for a harmonic function $U(r, s)$:

$$U(r,s) = \frac{1}{2\pi} \int_0^{2\pi} \frac{1-r^2}{1+r^2-2r\cos(t-s)} u(t)\, dt \,, \qquad (5)$$

inside the disk $r < 1$. The function $u(t) = U(1, t)$ is the prescribed value of the harmonic function on the circumference L of the disk. Set

$$z = re^{is} \,, \qquad \tau = e^{it}$$

in the relation (5) and get

$$U(r,s) = \text{Re}\left\{ \frac{1}{2\pi i} \int_L u(t) \frac{\tau+z}{\tau-z} \frac{d\tau}{\tau} \right\} . \qquad (6)$$

Now, let $V(r, s)$ be the function that is harmonic conjugate to $U(r, s)$:

$$U(r,s) + iV(r,s) = \frac{1}{2\pi i} \int_L u(t) \frac{\tau+z}{\tau-z} \frac{d\tau}{\tau} \,, \qquad (7)$$

such that $V(r, s)$ vanishes at the center of the disk:

$$V(r,s)\,|_{r=0} = 0 \,. \qquad (8)$$

Then, the function $V(r, s)$ is uniquely defined.

When $r \to 1$, so that z tends to a point ζ of the circumference L from within the disk, we can apply the Plemelj formula (8.4.2) to the analytic function (7). Therefore from (7), (8.4.2), and (4), we obtain

$$v(s) = -\frac{1}{2\pi} \int_0^{*2\pi} u(t) \cot\left(\frac{t-s}{2}\right) dt \,, \qquad (9)$$

where $v(s) = V(1, s)$ is the limiting value of the harmonic function on L. The formula (9) thus connects the limiting values of the conjugate harmonic functions $U(r, s)$ and $V(r, s)$ on the circumference.

We shall need the iterated formula formed by compounding the integrals with the Hilbert kernel:

$$g_1(s) = \frac{1}{2\pi} \int_0^{*2\pi} g(t) \cot\left(\frac{s-t}{2}\right) dt , \qquad (10)$$

and

$$g_2(s) = \frac{1}{2\pi} \int_0^{*2\pi} g_1(t) \cot\left(\frac{s-t}{2}\right) dt . \qquad (11)$$

If $U(r,s)$, $U_1(r,s)$, and $U_2(r,s)$ are the functions that are harmonic inside the disk $r < 1$, and whose values on the circumference $r = 1$ are equal respectively to $g(s)$, $g_1(s)$, and $g_2(s)$, then from (9) it follows that $U_1(r,s)$ is a harmonic function conjugate to $U(r,s)$ and $U_2(r,s)$ is conjugate to $U_1(r,s)$. The Cauchy–Riemann equations then yield the relations

$$\frac{\partial U}{\partial r} = -\frac{\partial U_2}{\partial r} , \qquad \frac{\partial U}{\partial s} = -\frac{\partial U_2}{\partial s} ,$$

or

$$U_2(r,s) = -U(r,s) + C , \qquad C = \text{const} . \qquad (12)$$

But the constant C can be determined from the condition (8):

$$C = U(r,s)\big|_{r=0} = (1/2\pi) \int_0^{2\pi} U(1,t) \, dt = (1/2\pi) \int_0^{2\pi} g(t) \, dt , \qquad (13)$$

where we have used the mean-value property of the harmonic function. Thereby, (12) becomes

$$U_2(r,s) = -U(r,s) + (1/2\pi) \int_0^{2\pi} g(t) \, dt . \qquad (14)$$

Now, put $r = 1$ in the above equation, use the relations $U(1,s) = g(s)$ and $U_2(1,s) = g_2(s)$, and get

$$g_2(s) = -g(s) + (1/2\pi) \int_0^{2\pi} g(t) \, dt . \qquad (15)$$

From (10), (11), and (15), we finally obtain the required iterated integral equation:

$$\left(\frac{1}{2\pi}\right)^2 \int_0^{*2\pi} \cot\left(\frac{\sigma - s}{2}\right) d\sigma \int_0^{*2\pi} g(t)\cot\left(\frac{t - \sigma}{2}\right) dt$$

$$= -g(s) + \frac{1}{2\pi}\int_0^{2\pi} g(t)\, dt\,, \tag{16}$$

which is called the Hilbert formula.

8.7. SOLUTION OF THE HILBERT-TYPE SINGULAR INTEGRAL EQUATION

We can solve the integral equation of the second kind

$$ag(s) = f(s) - \frac{b}{2\pi}\int_0^{*2\pi} g(t)\cot\left(\frac{t - s}{2}\right) dt\,, \tag{1}$$

where a and b are complex constants, in the same manner as we solved the corresponding Cauchy-type integral equation (8.5.1). As in Section 8.5, we define the operators L and M as

$$Lg = ag(s) + \frac{b}{2\pi}\int_0^{*2\pi} g(t)\cot\left(\frac{t - s}{2}\right) dt\,, \tag{2}$$

$$M\phi = a\phi(s) - \frac{b}{2\pi}\int_0^{*2\pi} \phi(t)\cot\left(\frac{t - s}{2}\right) dt\,. \tag{3}$$

Then,

$$MLg = a\left[ag(s) + \frac{b}{2\pi}\int_0^{*2\pi} g(t)\cot\left(\frac{t - s}{2}\right) dt\right] \qquad \text{(equation continued)}$$

$$-\frac{b}{2\pi}\int\limits_{0}^{*2\pi}\cot\left(\frac{t-s}{2}\right)dt$$

$$\times\left[ag(t)+\frac{b}{2\pi}\int\limits_{0}^{*2\pi}g(\sigma)\cot\left(\frac{\sigma-t}{2}\right)d\sigma\right]=F(s),\qquad(4)$$

where

$$F(s)=Mf=af(s)-\frac{b}{2\pi}\int\limits_{0}^{*2\pi}f(t)\cot\left(\frac{t-s}{2}\right)dt\ .$$

Using the Hilbert formula (8.6.16) and simplifying, we obtain the relation

$$(a^2+b^2)g(s)-(b^2/2\pi)\int\limits_{0}^{2\pi}g(t)\,dt=F(s)\ .\qquad(5)$$

The integral equation (5) has the simple degenerate kernel $K(s,t)=1$, and can be readily solved by the method of Chapter 2. The result is

$$g(s)=\frac{a}{a^2+b^2}f(s)-\frac{b}{2\pi(a^2+b^2)}\int\limits_{0}^{2\pi}f(t)\cot\left(\frac{t-s}{2}\right)dt$$

$$+\frac{b^2}{2\pi a(a^2+b^2)}\int\limits_{0}^{2\pi}f(t)\,dt\ ,\qquad(6)$$

provided $a^2+b^2\neq0$.

For the particular case $a=0$, the formula (6) is not applicable. Therefore, the solution of the Hilbert-type integral equation of the first kind

$$f(s)=\frac{b}{2\pi}\int\limits_{0}^{*2\pi}g(t)\cot\left(\frac{t-s}{2}\right)dt\ ,\qquad(7)$$

cannot be deduced from that of the second kind. But the integral equation (7) can be solved by other methods. For instance, the method of Section 8.1 is applicable here. Indeed, let us consider equation (7) with the constant b incorporated in $g(t)$:

$$f(s) = \frac{1}{2\pi} \int\limits_{0}^{*2\pi} g(t) \cot\left(\frac{t-s}{2}\right) dt . \tag{8}$$

In this equation, change s to t and t to σ, multiply both sides of the resulting equation by

$$\frac{1}{2\pi} \cot\left(\frac{t-s}{2}\right) ds ,$$

integrate from 0 to 2π, and use the Hilbert formula (8.6.16). There results the equation

$$g(s) - (1/2\pi) \int\limits_{0}^{2\pi} g(t)\, dt = F(s) , \tag{9}$$

where

$$F(s) = -\frac{1}{2\pi} \int\limits_{0}^{*2\pi} f(t) \cot\left(\frac{t-s}{2}\right) dt . \tag{10}$$

The integral equation (9) also has the simple kernel $K(s,t) = 1$ and therefore can be solved by the method of Chapter 2 if we set

$$(1/2\pi) \int\limits_{0}^{2\pi} g(s)\, ds = c .$$

Then, equation (9) becomes

$$g(s) - c = F(s) , \tag{11}$$

which, when integrated with respect to s from 0 to 2π, gives

$$\int\limits_{0}^{2\pi} F(s)\, ds = 0 . \tag{12}$$

From the relation (10), it follows that equation (12) holds for all values of the function $f(s)$. Therefore, the constant c is an arbitrary constant and the solution of the integral equation (8) is

$$g(s) = -\frac{1}{2\pi} \int\limits_{0}^{*2\pi} f(t) \cot\left(\frac{t-s}{2}\right) dt + c . \tag{13}$$

Finally, when we substitute (13) in (8), we find that the function $f(s)$ given by the relation (13) satisfies the original integral equation if, and only if,

$$\int_0^{2\pi} f(s) \, ds = 0 \, . \tag{14}$$

Hence, the condition (14) is necessary and sufficient for the Hilbert-type singular integral equation of the first kind to have a solution.

The second method is to use the results of Section 8.6, where we have connected the Hilbert kernel with the Cauchy kernel. For this purpose, we write $g(e^{it}) = g(t)$, etc., and assume that $g(t)$ and $f(t)$ are periodic functions with period 2π. Further, we replace $f(t)$ by $f(t)/i$. Then, the formulas (8.5.8) with the help of the transformation (8.6.3) yield the reciprocal relations

$$\frac{1}{2\pi} {\int_0^{2\pi}}^{*} g(t) \cot\left(\frac{t-s}{2}\right) dt + \frac{i}{2\pi} \int_0^{2\pi} g(t) \, dt = f(s) \tag{15}$$

and

$$\frac{1}{2\pi} {\int_0^{2\pi}}^{*} f(t) \cot\left(\frac{t-s}{2}\right) dt + \frac{i}{2\pi} \int_0^{2\pi} f(t) \, dt = -g(s) \, . \tag{16}$$

With the help of this pair of equations, the solution of the integral equation (8) can be easily deduced.

It follows from the pair (15)–(16) that, for periodic functions $f(t)$ and $g(t)$, if the condition $\int_0^{2\pi} f(t) \, dt = 0$ is satisfied, then we also have $\int_0^{2\pi} g(t) \, dt = 0$. We shall have more to say about the integral relations (15) and (16) in the next chapter.

8.8. EXAMPLES

Example 1. Prove that the integral equation

$$F(v) = \int_0^{*1} wf(w) \frac{e^{-i\alpha(w-v)}}{w-v} \, dw \qquad (1)$$

has the solution

$$wf(w) = -\frac{1}{\pi^2} \left(\frac{w}{1-w}\right)^{\frac{1}{2}} \int_0^{*1} \left(\frac{1-v}{v}\right)^{\frac{1}{2}} F(v) \frac{e^{-i\alpha(v-w)}}{v-w} \, dv. \qquad (2)$$

The integral equation (1) arises in the discussion of various problems of mathematical physics. The solution follows by comparing (1) with (8.5.18) and (2) with (8.5.19), and by setting $\alpha = 0$, $\beta = 1$.

Example 2.   Solve the integral equation

$$\sum_1^\infty (a_n \cos ns + b_n \sin ns) = \frac{1}{2\pi} \int_0^{*2\pi} g(t) \cot\left(\frac{t-s}{2}\right) dt . \qquad (3)$$

Observe that the function $f(s) = \sum_1^\infty (a_n \cos ns + b_n \sin ns)$ is a periodic function with period 2π. Moreover, the condition $\int_0^{2\pi} f(t) \, dt = 0$ is satisfied. Therefore, from the reciprocal pair (8.7.15) and (8.7.16), it follows that

$$g(s) = -\frac{1}{2\pi} \int_0^{*2\pi} \left[\sum_{n=1}^\infty (a_n \cos nt + b_n \sin nt)\right] \cot\left(\frac{t-s}{2}\right) dt$$

$$= \sum_{n=1}^\infty (a_n \sin ns - b_n \cos ns) , \qquad (4)$$

where we have left the actual integration as an exercise for the reader.

EXERCISES

Solve the following integral equations.

1. $a + bs + cs^2 + ds^3 = \int_1^s \dfrac{g(t) \, dt}{(\cos t - \cos s)^{\frac{1}{2}}}$, $1 < s < 2$, where a, b, c, d are real constants.

2. $a + bs + cs^2 + ds^3 = \int\limits_{s}^{2} \dfrac{g(t)\, dt}{(\cos s - \cos t)^{\frac{1}{2}}},$ $1 < s < 2.$

3. $s^2 = \int\limits_{2}^{s} \dfrac{g(t)\, dt}{(s^2 - t^2)^{\frac{1}{3}}},$ $2 < s < 4.$

4. $s^2 = \int\limits_{s}^{4} \dfrac{g(t)\, dt}{(t^2 - s^2)^{\frac{1}{3}}},$ $2 < s < 4.$

5. $as + bs^2 = \int\limits_{0}^{s} \dfrac{g(t)\, dt}{(s - t)^{\frac{1}{2}}}.$

6. Substitute the solution (8.5.5) in (8.5.1) and verify that this solution satisfies the given integral equation.

7. Prove that the solution of the integral equation (8.7.5) is (8.7.6).

8. Find the solution of the integral equation

$$g(s) = (\sin s) - \frac{1}{2\pi} \int\limits_{0}^{*2\pi} g(t) \cot\left(\frac{t-s}{2}\right) dt .$$

9. Solve the integral equation

$$s^{-1} \int\limits_{0}^{s} \frac{t^2 g(t)\, dt}{(s^2 - t^2)^{\frac{1}{2}}} = \begin{cases} -f_1(s), & 0 < s \leqslant b, \\ f_2(s), & b < s < \infty, \end{cases}$$

where b is a constant.

10. Prove that the solution of the integral equation

$$f(s) = \frac{2s^{-2\alpha - 2\eta}}{\Gamma(\alpha)} \int\limits_{0}^{s} t^{2\eta + 1} (s^2 - t^2)^{\alpha - 1} g(t)\, dt , \qquad 0 < \alpha < 1,$$

is

$$g(t) = \frac{t^{-2\eta-1}}{\Gamma(1-\alpha)} \frac{d}{dt} \int_0^t u^{2\alpha+2\eta+1}(t^2-u^2)^{-\alpha}f(u)\, du \ .$$

11. Prove that the solution of the integral equation

$$f(s) = \frac{2s^{2\eta}}{\Gamma(\alpha)} \int_s^\infty (t^2-s^2)^{\alpha-1}\, t^{-2\alpha-2\eta+1}\, g(t)\, dt \ , \qquad 0 < \alpha < 1 \ ,$$

is

$$g(t) = -\frac{t^{2\alpha+2\eta-1}}{\Gamma(1-\alpha)} \frac{d}{dt} \int_t^\infty u^{-2\eta+1}(u^2-t^2)^{-\alpha}f(u)\, du \ .$$

12. Solve the integral equation

$$s \int_s^\infty \frac{g(t)\, dt}{(t^2-s^2)^{\frac{1}{2}}} = \begin{cases} f_1(s), & 0 \leqslant s \leqslant a \ , \\ -f_2(s), & a < s < \infty \ . \end{cases}$$

INTEGRAL TRANSFORM METHODS

CHAPTER 9

9.1. INTRODUCTION

The integral transform methods are of great value in the treatment of integral equations, especially the singular integral equations. Suppose that a relationship of the form

$$g(s) = \iint \Gamma(s, x) K(x, t) g(t) \, dt \, dx \qquad (1)$$

is known to be valid and that this double integral can be evaluated as an iterated integral. This means that the solution of the integral equation of the first kind,

$$f(s) = \int K(s, t) g(t) \, dt , \qquad (2)$$

is

$$g(s) = \int \Gamma(s, t) f(t) \, dt . \qquad (3)$$

Conversely, the relation (2) can be considered as the solution of the integral equation (3). It is conventional to refer to one of these functions as the transform of the second and to the second as an inverse transform of the first.

194

The most celebrated example of the double integral (1) is the Fourier integral

$$g(s) = (1/2\pi) \int_{-\infty}^{\infty} \int_{-\infty}^{\infty} e^{isx} e^{-ixt} g(t) \, dt \, dx \,, \tag{4}$$

which results in the reciprocal relations

$$f(s) = (2\pi)^{-\frac{1}{2}} \int_{-\infty}^{\infty} e^{-ist} g(t) \, dt \tag{5}$$

and

$$g(s) = (2\pi)^{-\frac{1}{2}} \int_{-\infty}^{\infty} e^{ist} f(t) \, dt \,. \tag{6}$$

The function $f(s)$ is known as the Fourier transform $T[g]$ of $g(t)$ and $g(s)$ as the inverse transform $T^{-1}[f]$ of $f(s)$, and vice versa. The function $f(s)$ exists if $g(t)$ is absolutely integrable, and it is square-integrable if $g(t)$ is square-integrable, as can be readily verified using Bessel's inequality. In the sequel, we shall assume that the functions involved in the integral equations as well as their transforms satisfy the appropriate regularity conditions, so that the required operations are valid.

As a second example, consider the double integral

$$g(s) = (2/\pi) \int_{0}^{\infty} \int_{0}^{\infty} (\sin sx \sin xt) g(t) \, dt \, dx \,. \tag{7}$$

This leads to the sine transform and its inverse,

$$f(s) = (2/\pi)^{\frac{1}{2}} \int_{0}^{\infty} (\sin st) g(t) \, dt \tag{8}$$

and

$$g(s) = (2/\pi)^{\frac{1}{2}} \int_{0}^{\infty} (\sin st) f(t) \, dt \,, \tag{9}$$

respectively.

For ease of notation, we shall also call the transform of f as F and that of g as G, etc., for all the transforms. It will be clear in the context as to what transforms we are implying.

9.2. FOURIER TRANSFORM

The Fourier transform $T[f]$,

$$T[f] = F(s) = (2\pi)^{-\frac{1}{2}} \int_{-\infty}^{\infty} f(t) e^{-ist} dt \,, \tag{1}$$

is a linear transformation:

$$T[af+bg] = (2\pi)^{-\frac{1}{2}} \int_{-\infty}^{\infty} [af(t)+bg(t)] e^{-ist} dt$$

$$= a(2\pi)^{-\frac{1}{2}} \int_{-\infty}^{\infty} f(t) e^{-ist} dt + b(2\pi)^{-\frac{1}{2}} \int_{-\infty}^{\infty} g(t) e^{-ist} dt$$

$$= aT[f] + bT[g] \,. \tag{2}$$

As such, we can use many properties of the linear operators. Furthermore, in Chapter 7 (see Exercise 11), we found that under Fourier transformation a square-integrable function preserves its norm. Hence, for such a function, we have

$$\|T[f]\| = \|F\| = \|f\| \,. \tag{3}$$

Let us note some of the important properties of the Fourier transforms. They can be found in every standard book on the subject (see, for example, [17]) and, in fact, can be proved very easily by the mere use of the definitions above. These properties are:

(i) $T[f(t-a)] = e^{-ias} T[f(t)] \,,$ (4)

where a is a constant.

(ii) $T[f(at)] = (1/|a|) T[f(t)]_{s \to s/a} \,.$ (5)

(iii) $T[f'(t)] = isT[f(t)] \,,$ (6)

where the prime denotes differentiation with respect to the argument. Similarly,

$$T[f^k(t)] = (is)^k T[f(t)] \,, \tag{7}$$

where by $f^k(t)$ we mean the kth derivative of f.

(iv)   If

$$h(t) = \int_a^t f(x)\, dx \,, \tag{8}$$

then

$$T[h(t)] = (1/is)\, T[f(t)] \,. \tag{9}$$

From (6) and (9), we see that the differentiation has the effect of multiplying the transform by is, whereas integration has the effect of dividing the transform by is.

(v)   The convolution integral

$$h(t) = (2\pi)^{-\frac{1}{2}} \int_{-\infty}^{\infty} f(t-x)\, g(x)\, dx = (2\pi)^{-\frac{1}{2}} \int_{-\infty}^{\infty} g(t-x)f(x)\, dx \tag{10}$$

gives

$$T[h(t)] = T[f]\, T[g] \,, \tag{11}$$

or

$$H(s) = F(s)\, G(s) \,. \tag{12}$$

9.3. LAPLACE TRANSFORM

The Laplace transform $L[f]$ of a function $f(s)$ is defined as

$$L[f] = F(p) = \int_0^\infty f(s)\, e^{-ps}\, ds \,. \tag{1}$$

The inverse $L^{-1}[F]$ is

$$L^{-1}[F] = f(s) = (1/2\pi i) \int_{\gamma-i\infty}^{\gamma+i\infty} F(p)\, e^{ps}\, dp \,. \tag{2}$$

This transformation is also linear because

$$L[af+bg] = aL[f] + bL[g] \,,$$

for two constants a and b.

The following are some of the basic properties of the Laplace transform:

(i) $F(p-a) = L[e^{as}f(s)]$,   (3)

(ii) $L[f(as)] = (1/a)L[f(s)]_{p \to p/a}$,   (4)

(iiia) $L[f'] = pL[f] - f(0)$,   (5)

(b) $L[f^k] = p^k L[f] - p^{k-1}f(0) - p^{k-2}f'(0) - \cdots - f^{k-1}(0)$,   (6)

(c) $dF(p)/dp = -L[sf(s)]$,   (7)

where f^k means kth derivative with respect to the argument.

(iv)   If

$$h(s) = \int_0^s f(x)\, dx ,$$ (8)

then

$$L[h] = H(p) = (1/p)L[f] .$$ (9)

(v)   For the convolution integral

$$h(s) = \int_0^s f(x)\,g(s-x)\, dx = \int_0^s g(x)f(s-x)\, dx ,$$ (10)

we have

$$L[h] = L[f]L[g] ,$$ (11)

or

$$H(p) = F(p)\,G(p) ,$$ (12)

which is the same as (9.2.12).

9.4. APPLICATIONS TO VOLTERRA INTEGRAL EQUATIONS WITH CONVOLUTION-TYPE KERNELS

The basic information given here about the Fourier and Laplace transforms is sufficient to demonstrate their application to the solution of integral equations. We shall apply only the Laplace transform in this section, although the Fourier transform can also be applied just as effectively. Let us first consider the Volterra-type integral equation of the first kind,

$$f(s) = \int_0^s k(s-t)\,g(t)\,dt \,, \tag{1}$$

where $k(s-t)$ depends only on the difference $(s-t)$. Applying the Laplace transform to both sides of this equation, we obtain

$$F(p) = K(p)\,G(p)$$

or

$$G(p) = F(p)/K(p) \,. \tag{2}$$

The solution follows by inversion.

The present method is also applicable to the Volterra integral equation of the second kind with a convolution-type kernel

$$g(s) = f(s) + \int_0^s k(s-t)\,g(t)\,dt \,. \tag{3}$$

On applying Laplace transformation to both sides and using the convolution formula, we have

$$G(p) = F(p) + K(p)\,G(p)$$

or

$$G(p) = F(p)/[1 - K(p)] \,, \tag{4}$$

and inversion yields the solution.

We can also find the resolvent kernel of the integral equation (3) by integral transform methods. For this purpose, we first show that, if the original kernel $k(s,t)$ is a difference kernel, then so is the resolvent kernel. Since the resolvent kernel $\Gamma(s,t)$ is a sum of the iterated kernels, all that we have to prove is that they all depend on the difference $(s-t)$. Indeed,

$$k_2(s,t) = \int_t^s k(s-x)\,k(x-t)\,dx = \int_0^{s-t} k(s-t-\sigma)\,k(\sigma)\,d\sigma \,, \tag{5}$$

where we have set $\sigma = x - t$. This process can obviously be continued and our assertion is proved. Hence, the solution of the integral equation (3) is

$$g(s) = f(s) + \int_0^s \Gamma(s-t) f(t) \, dt \ . \tag{6}$$

Application of the Laplace transform to both sides of (6) gives

$$G(p) = F(p) + \Omega(p) F(p) \ , \tag{7}$$

where

$$\Omega(p) = L[\Gamma(s-t)] \ . \tag{8}$$

From (4) and (7), we have

$$F(p)/[1 - K(p)] = F(p)[1 + \Omega(P)] \ , \tag{9}$$

$$\Omega(p) = K(p)/[1 - K(p)] \ . \tag{10}$$

By inversion, we recover $\Gamma(s-t)$.

We illustrate the above ideas with numerous examples. Throughout these examples, we have left the evaluation of the Laplace transform and its inverse to the reader. There are numerous monographs that contain the required formulas [3, 12, 17].

9.5. EXAMPLES

Example 1.   Solve the Abel integral equation

$$f(s) = \int_0^s [g(t)/(s-t)^\alpha] \, dt \ , \qquad 0 < \alpha < 1 \ . \tag{1}$$

This is a convolution integral and therefore

$$F(p) = K(p) G(p) \ , \tag{2}$$

where $K(p)$ is the Laplace transform of $k(s) = s^{-\alpha}$:

$$K(p) = p^{\alpha - 1} \Gamma(1 - \alpha) \ . \tag{3}$$

From (2) and (3), it follows that

$$G(p) = \frac{p^{1-\alpha} F(p)}{\Gamma(1-\alpha)} = \frac{p}{\Gamma(\alpha)\Gamma(1-\alpha)} \{\Gamma(\alpha)p^{-\alpha} F(p)\}$$

$$= \frac{p}{\pi \csc \pi\alpha} \{\Gamma(\alpha)p^{-\alpha} F(p)\} ,\tag{4}$$

where we have used the relation $\Gamma(\alpha)\Gamma(1-\alpha) = \pi \csc \pi\alpha$. Now if we use the relation (9.3.12), (4) becomes

$$G(p) = \frac{\sin \alpha\pi}{\pi} pL\left[\int_0^s (s-t)^{\alpha-1} f(t)\, dt \right].\tag{5}$$

By virtue of the property (9.3.5), we finally have

$$g(s) = \frac{\sin \alpha\pi}{\pi} \frac{d}{ds} \int_0^s (s-t)^{\alpha-1} f(t)\, dt ,\tag{6}$$

which agrees with the relation (8.1.4) obtained in the previous chapter by a different method.

Example 2. Solve the integral equation

$$s = \int_0^s e^{s-t} g(t)\, dt .\tag{7}$$

Taking the Laplace transform of both sides, we obtain

$$1/p^2 = K(p)G(p) ,\tag{8}$$

where $K(p)$ is the Laplace transform of $k(s) = e^s$:

$$K(p) = \int_0^\infty e^s e^{-sp}\, ds = 1/(p-1) .\tag{9}$$

The result of combining (7), (8), and (9) is

$$G(p) = (p-1)/p^2 = (1/p) - (1/p^2) ,$$

whose inverse is

$$g(s) = 1 - s . \tag{10}$$

Example 3.   Solve the integral equation

$$\sin s = \int_0^s J_0(s-t) g(t) \, dt . \tag{11}$$

Here, the function $k(s) = J_0(s)$, whose Laplace transform is known to be $1/(1+p^2)^{1/2}$. Also, the Laplace transform of $\sin s$ is $1/(1+p^2)$. Therefore, when we take the Laplace transform of equation (11), there results the relation

$$G(p) = 1/(1+p^2)^{1/2} ,$$

which by inversion yields the solution

$$g(s) = J_0(s) . \tag{12}$$

Incidently, by substituting (12) back in (11), we get the interesting result

$$\int_0^s J_0(s-t) J_0(t) \, dt = \sin s . \tag{13}$$

Example 4.   Recall that we solved integral equations of the type

$$f(s) = \int_0^s k(s^2 - t^2) g(t) \, dt , \qquad s > 0 , \tag{14}$$

in the previous chapter for some special cases of the kernel $k(s^2 - t^2)$. With the help of the Laplace transform, we can solve equation (14) for a general convolution kernel. For this purpose, the first step is to set

$$s = u^{1/2} , \qquad t = \sigma^{1/2} , \qquad g_1(\sigma) = \tfrac{1}{2}\sigma^{-1/2} g(\sigma^{1/2}) , \qquad f_1(u) = f(u^{1/2}) . \tag{15}$$

Then, the integral equation (14) takes the form

$$f_1(u) = \int_0^u k(u-\sigma) g_1(\sigma) \, d\sigma , \qquad u > 0 . \tag{16}$$

Taking the Laplace transform of both sides of (16), we get

$$G_1(p) = F_1(p)/K(p) = pF_1(p)/pK(p) . \tag{17}$$

By defining

$$1/pK(p) = H(p) , \tag{18}$$

the relation (17) becomes

$$G_1(p) = pH(p)F_1(p) . \tag{19}$$

Using the relations (9.3.5) and (9.3.12), we have

$$G_1(p) = L\left[\frac{d}{du}\int_0^u h(u-\sigma)f_1(\sigma)\,d\sigma\right] \tag{20}$$

or

$$g_1(u) = \frac{d}{du}\int_0^u h(u-\sigma)f_1(\sigma)\,d\sigma , \tag{21}$$

where $h(s)$ stands for the inverse of the function $H(p)$. Finally, from (15) and (21), we have the required solution

$$g(s) = 2\frac{d}{ds}\int_0^s tf(t)h(s^2-t^2)\,dt . \tag{22}$$

Let us solve (14) for some special cases:

(a) $k(t) = t^{-\alpha}$, $0 < \alpha < 1$. This means that we have to solve the integral equation

$$f(s) = \int_0^s [g(t)/(s^2-t^2)^\alpha]\,dt . \tag{23}$$

The solution follows from (22) if we can evaluate the function $h(s)$ from the relation

$$H(p) = 1/pK(p) . \tag{24}$$

But

$$K(p) = \int_0^\infty t^{-\alpha} e^{-pt} \, dt = \Gamma(1-\alpha) p^{\alpha-1} \,.$$

Therefore,

$$h(s) = L^{-1}\left[\frac{1}{p^\alpha \Gamma(1-\alpha)}\right] = \frac{\sin \alpha \pi}{\pi} s^{\alpha-1} \,,$$

and

$$g(s) = \frac{2 \sin \alpha \pi}{\pi} \frac{d}{ds} \int_0^s \frac{t f(t) \, dt}{(s^2 - t^2)^{1-\alpha}} \,, \tag{25}$$

which agrees with the relation (8.2.9).

(b) $k(t) = t^{-\frac{1}{2}} \cos(\beta t^{\frac{1}{2}})$, where β is a constant. In this case, $K(p) = \pi^{\frac{1}{2}} p^{-\frac{1}{2}} \exp(-\beta^2/4p)$. Therefore,

$$h(s) = L^{-1}[\pi^{-\frac{1}{2}} p^{-\frac{1}{2}} \exp(\beta^2/4p)] = \pi^{-1} s^{-\frac{1}{2}} \cosh(\beta t^{\frac{1}{2}}) \,. \tag{26}$$

This means that the solution of the integral equation

$$f(s) = \int_0^s \frac{\cos[\beta(s^2 - t^2)^{\frac{1}{2}}]}{(s^2 - t^2)^{\frac{1}{2}}} g(t) \, dt \,, \qquad s > 0 \,, \tag{27}$$

is

$$g(s) = \frac{2}{\pi} \frac{d}{ds} \int_0^s \frac{\cosh[\beta(s^2 - t^2)^{\frac{1}{2}}]}{(s^2 - t^2)^{\frac{1}{2}}} t f(t) \, dt \,. \tag{28}$$

Note that the relations (27) and (28) remain valid for $0 < s < \infty$.

Example 5. Solve the inhomogeneous integral equation

$$g(s) = 1 - \int_0^s (s-t) g(t) \, dt \,. \tag{29}$$

Since

$$k(s) = s \,, \qquad K(p) = 1/p^2 \,, \tag{30}$$

the application of the Laplace transform gives

$$G(p) = (1/p) - [G(p)/p^2]$$

or

$$G(p) = p/(1+p^2) = L[\cos s] .$$

Hence, the solution is

$$g(s) = \cos s . \tag{31}$$

Example 6. Find the resolvent of the integral equation

$$g(s) = f(s) + \int_0^s (s-t)g(t) \, dt . \tag{32}$$

Here again $k(s) = s$, and we have $K(p) = 1/p^2$. The formula (9.4.10) gives $\Omega(p) = 1/(p^2 - 1)$, whose inverse is $\Gamma(s) = \frac{1}{2}(e^s - e^{-s})$. Therefore, the value of the resolvent kernel is

$$\Gamma(s-t) = \frac{1}{2}(e^{s-t} - e^{-s+t}) ,$$

and the solution of the integral equation (32) is

$$g(s) = f(s) + \frac{1}{2}e^s \int_0^s e^{-t}f(t) \, dt - \frac{1}{2}e^{-s} \int_0^s e^t f(t) \, dt . \tag{33}$$

Example 7. Find the resolvent of the integral equation

$$g(s) = f(s) + \int_0^s e^{s-t}g(t) \, dt . \tag{34}$$

Here, $k(s) = e^s$, which gives $K(p) = 1/(p-1)$, and from the formula (9.4.10), we have $\Omega(p) = 1/(p-2)$, $\Gamma(s) = e^{2s}$. Hence, the resolvent kernel is $\Gamma(s-t) = e^{2s-2t}$, and the solution of the integral equation (34) is

$$g(s) = f(s) + \int_0^s e^{2s-2t}f(t) \, dt . \tag{35}$$

Example 8. Solve the integral equation

$$g(s) = f(s) + \lambda \int_0^s J_0(s-t)g(t) \, dt . \tag{36}$$

The kernel $k(s) = \lambda J_0(s)$, and therefore

$$K(p) = \frac{\lambda}{(1+p^2)^{\frac{1}{2}}}, \qquad \Omega(p) = \frac{\lambda}{(1+p^2)^{\frac{1}{2}} - \lambda},$$

$$\Gamma(s) = \frac{\lambda}{(1-\lambda^2)^{\frac{1}{2}}} \int_0^s [\sin(1-\lambda^2)^{\frac{1}{2}}](s-\sigma)\frac{J_1(\sigma)}{\sigma} \, d\sigma$$

$$+ \lambda \{\cos[(1-\lambda^2)^{\frac{1}{2}} s]\} + \frac{\lambda^2}{(1-\lambda^2)^{\frac{1}{2}}} \sin[(1-\lambda^2)^{\frac{1}{2}} s] \,. \tag{37}$$

The value of the resolvent kernel follows by setting $(s-t)$ for s and the solution of the integral equation (35) is then readily obtained from the formula (9.4.6).

Example 9. As a final example, solve the inhomogeneous Abel integral equation:

$$g(s) = f(s) + \lambda \int_0^s [g(t)/(s-t)^\alpha] \, dt \,, \qquad 0 < \alpha < 1 \,. \tag{38}$$

The kernel $k(s) = \lambda s^{-\alpha}$ yields

$$K(p) = \lambda \Gamma(1-\alpha) p^{\alpha-1} \,,$$
$$\Omega(p) = \lambda \Gamma(1-\alpha) p^{\alpha-1}/[1 - \lambda \Gamma(1-\alpha) p^{\alpha-1}] \,. \tag{39}$$

The inverse of (39) is

$$\Gamma(s) = \sum_{n=1}^\infty \frac{[\lambda \Gamma(1-\alpha) s^{1-\alpha}]^n}{s \Gamma[n(1-\alpha)]} \,. \tag{40}$$

Hence, the solution of the integral equation (38) is

$$g(s) = f(s) + \int_0^s \frac{\sum_{n=1}^\infty [\lambda \Gamma(1-\alpha)(s-t)^{1-\alpha}]^n}{(s-t)\Gamma[n(1-\alpha)]} f(t) \, dt \,. \tag{41}$$

9.6. HILBERT TRANSFORM

The finite Hilbert transform of a function $g(\varphi)$ is usually defined as

$$f(\theta) = \frac{1}{\pi} \int_0^{*\pi} \frac{\sin\theta}{\cos\theta - \cos\varphi} \, g(\varphi) \, d\varphi \, , \tag{1}$$

with the inverse

$$g(\theta) = \frac{1}{\pi} \int_0^{*\pi} \frac{\sin\varphi}{\cos\varphi - \cos\theta} f(\varphi) \, d\varphi + \frac{1}{\pi} \int_0^{\pi} g(\varphi) \, d\varphi \, . \tag{2}$$

Various other forms of the Hilbert transform pair can be deduced from equations (1) and (2). In this connection, we need the relation

$$\int_0^{*\pi} \frac{\cos n\varphi \, d\varphi}{\cos\varphi - \cos\alpha} = \pi \frac{\sin n\alpha}{\sin\alpha} \, , \tag{3}$$

which can be proved by induction since the cases $n = 0$ and $n = 1$, are elementary relations.

From (1) and (2), it is apparent that $f(-\theta) = -f(\theta)$, $g(-\theta) = g(\theta)$. Now, set

$$f_1(\theta) = -f(\theta) \, , \qquad \text{so that} \quad f_1(-\theta) = -f_1(\theta) \, , \tag{4}$$

$$(1/\pi) \int_0^{\pi} g(\varphi) \, d\varphi = C \, , \qquad C = \text{const} \, , \tag{5}$$

$$g_1(\theta) = g(\theta) - C \, , \qquad \text{so that} \quad g_1(-\theta) = g(\theta) - C \, . \tag{6}$$

From the above relations, it follows that

$$(1/\pi) \int_0^{\pi} g_1(\theta) \, d\theta = (1/\pi) \int_0^{\pi} g(\theta) \, d\theta - (1/\pi) \int_0^{\pi} C \, d\theta = 0 \, , \tag{7}$$

and

$$\frac{1}{\pi} \int_0^{*\pi} \frac{\sin\theta}{\cos\varphi - \cos\theta} \, g_1(\varphi) \, d\varphi = \frac{1}{\pi} \int_0^{*\pi} \frac{\sin\theta}{\cos\varphi - \cos\theta} \, [g(\varphi) - C] \, d\varphi$$

$$= -f(\theta) - \frac{C}{\pi} \int\limits_0^{*\pi} \frac{\sin\theta}{\cos\varphi - \cos\theta}\, d\varphi = f_1(\theta)\,.$$

$$(8)$$

Writing $\theta = \frac{1}{2}(\theta - \varphi) + \frac{1}{2}(\theta + \varphi)$ in equation (8), we get

$$f_1(\theta) = (1/2\pi) \int\limits_0^{*\pi} [\cot\tfrac{1}{2}(\theta + \varphi)]\, g_1(\varphi)\, d\varphi$$

$$+ (1/2\pi) \int\limits_0^{*\pi} [\cot\tfrac{1}{2}(\theta - \varphi)]\, g_1(\varphi)\, d\varphi\,. \qquad (9)$$

In the first integral, replace φ by $-\varphi$ and use the relation (6). Combine the resulting integral with the second integral in (9). The result is

$$f_1(\theta) = (1/2\pi) \int\limits_{-\pi}^{*\pi} [\cot\tfrac{1}{2}(\theta - \varphi)]\, g_1(\varphi)\, d\varphi\,. \qquad (10)$$

Similarly, by starting with equation (2) and going through the same steps as above, we get the relation

$$g_1(\theta) = (1/2\pi) \int\limits_0^{*\pi} [\cot\tfrac{1}{2}(\varphi - \theta)]\, f_1(\varphi)\, d\varphi$$

$$+ (1/2\pi) \int\limits_0^{*\pi} [\cot\tfrac{1}{2}(\varphi + \theta)]\, f_1(\varphi)\, d\varphi\,. \qquad (11)$$

After replacing φ by $-\varphi$ in the second integral and using the relation (4), we obtain from (11)

$$g_1(\theta) = (1/2\pi) \int\limits_{-\pi}^{*\pi} [\cot\tfrac{1}{2}(\varphi - \theta)]\, f_1(\varphi)\, d\varphi\,. \qquad (12)$$

The relations (10) and (12) constitute a second form of the finite Hilbert transform pair.

The third form of the Hilbert transform pair can be deduced from (9) and (11). In fact, the relation (9) can be written as

$$f_1(\theta) = (1/2\pi) \int\limits_0^{*\pi} [1 + \cot\tfrac{1}{2}(\theta + \varphi)]\, g_1(\varphi)\, d\varphi$$

$$+ (1/2\pi) \int_0^{*\pi} [1 + \cot\tfrac{1}{2}(\theta - \varphi)] g_1(\varphi)\, d\varphi - (1/\pi) \int_0^{\pi} g_1(\varphi)\, d\varphi .$$

$$(13)$$

The last integral vanishes because of (7), while the first integral can be combined with the second by replacing φ by $-\varphi$ and using (6). The result is

$$f_1(\theta) = (1/2\pi) \int_{-\pi}^{*\pi} [1 + \cot\tfrac{1}{2}(\theta - \varphi)] g_1(\varphi)\, d\varphi . \qquad (14)$$

Similarly, the relation (11) takes the form

$$g_1(\theta) = (1/2\pi) \int_{-\pi}^{*\pi} [1 + \cot\tfrac{1}{2}(\varphi - \theta)] f_1(\varphi)\, d\varphi . \qquad (15)$$

The transform pair (14)–(15) is precisely the reciprocal pair of Hilbert-type singular integral equations encountered in Section 8.7 of the previous chapter except for a trivial adjustment of the symbols and the range of integration.

A fourth form of the finite Hilbert transform pair, which is nonangular, is obtained from the pair (1)–(2) by setting

$$x = \cos\theta, \qquad y = \cos\varphi, \qquad p(x) = \frac{f(\theta)}{\sin\theta} = \frac{f(\cos^{-1} x)}{(1 - x^2)^{1/2}},$$

$$q(x) = \frac{g(\theta)}{\sin\theta} = \frac{g(\cos^{-1} x)}{(1 - x^2)^{1/2}} . \qquad (16)$$

Then, equation (1) becomes

$$\frac{f(\theta)}{\sin\theta} = \frac{1}{\pi} \int_0^{*\pi} \frac{1}{\cos\theta - \cos\varphi} \frac{g(\varphi)}{\sin\varphi} \sin\varphi\, d\varphi$$

or

$$p(x) = \frac{1}{\pi} \int_{-1}^{*1} \frac{q(y)\, dy}{x - y}, \qquad -1 < x < 1, \qquad (17)$$

the so-called airfoil equation. Similarly, (2) takes the form

$$q(x) = \frac{1}{\pi} \int_{-1}^{*\,1} \left(\frac{1-y^2}{1-x^2}\right)^{\frac{1}{2}} \frac{p(y)}{y-x} \, dy + \frac{C}{(1-x^2)^{\frac{1}{2}}}, \qquad -1 < x < 1, \quad (18)$$

where

$$C = (1/\pi) \int_{-1}^{1} q(y) \, dy$$

has the character of an arbitrary constant.

The pair of equations (17)–(18) is a special case of the pair of integral equations (8.5.18)–(8.5.19).

The infinite Hilbert transform is defined as

$$f(s) = (1/\pi) \int_{-\infty}^{*\,\infty} [g(t)/(t-s)] \, dt . \tag{19}$$

Its inverse is

$$g(s) = -(1/\pi) \int_{-\infty}^{*\,\infty} [f(t)/(t-s)] \, dt . \tag{20}$$

9.7. EXAMPLES

Example 1.   Solve the homogeneous integral equation

$$\int_{-1}^{*\,1} [g(y)/(x-y)] \, dy = 0 . \tag{1}$$

The solution follows from (9.6.18):

$$g(x) = C/(1-x^2)^{\frac{1}{2}} . \tag{2}$$

Example 2.   Solve the integral equation

$$\sin s = (1/\pi) \int_{-\infty}^{*\,\infty} [g(t)/(t-s)] \, dt . \tag{3}$$

To solve this equation, let us consider the integral

$$\int_{-\infty}^{*\infty} [e^{it}/(s-t)]\, dt = \pi i \sum \text{(residues of the poles on the } t \text{ axis)}$$

$$= \pi i(\cos s + i \sin s) . \tag{4}$$

Separating the real and imaginary parts, we obtain

$$(1/\pi) \int_{-\infty}^{*\infty} [(\cos t)/(s-t)]\, dt = -\sin s , \tag{5}$$

$$(1/\pi) \int_{-\infty}^{*\infty} [(\sin t)/(s-t)]\, dt = \cos s . \tag{6}$$

Comparing (3) and (5), we have the solution

$$g(t) = \cos t . \tag{7}$$

EXERCISES

1. Show that the solution of the integral equation

$$f(s) = 2 \int_{s}^{1} \frac{t g(t)\, dt}{(t^2 - s^2)^{1/2}}$$

is

$$g(s) = -\frac{1}{\pi s} \frac{d}{ds} \int_{s}^{1} \frac{t f(t)\, dt}{(t^2 - s^2)^{1/2}} .$$

Find the solution for the following two special cases: (i) $f(s) = 2s^2/(1-s^2)^{1/2}$; (ii) $f(s) = s^2$.

2. Solve the integral equation

$$f(s) = s \int_{s}^{\infty} \frac{g'(t)}{(t^2 - s^2)^{1/2}}\, dt .$$

3. Solve the Abel integral equation of the second kind

$$g(s) = s^{-\frac{1}{2}} e^{-a/4s} + \frac{i}{\sqrt{\pi}} \int_0^s \frac{g(t)}{(s-t)^{\frac{1}{2}}} \, dt \ .$$

This integral equation arises in the theory of wave propagation over a flat surface.

4. If it is required that in the Hilbert transform pair (9.6.17)–(9.6.18) the function $q(-1)$ be finite, show that there must follow

$$\int_{-1}^{1} q(t) \, dt = \int_{-1}^{1} (1-t^2)^{\frac{1}{2}} p(t) \frac{dt}{1+t} \ ,$$

and verify that in this case the solution $q(s)$ becomes

$$q(s) = \frac{1}{\pi} \left(\frac{1+s}{1-s} \right)^{\frac{1}{2}} \int_{-1}^{1} \left(\frac{1-t}{1+t} \right)^{\frac{1}{2}} \frac{p(t)}{s-t} \, dt \ .$$

5. With the help of finite Hilbert transform, solve the equation

$$s^2 = \int_{-\ell}^{\ell} \frac{2tg(t)}{s^2 - t^2} \, dt \ , \qquad \text{assuming that} \qquad g(t) = -g(-t).$$

6. With the help of finite Hilbert transform, solve the equation

$$as + b + \sigma_\ell (\log |\ell - s|) - \sigma_0 \log |s| = \int_0^\ell [g'(t)/(t-s)] \, dt$$

subject to the conditions

$$g'(0) = \sigma_0 \ , \qquad g'(\ell) = \sigma_\ell \ , \qquad g(0) = g_0 \ , \qquad g(\ell) = g_\ell \ .$$

7. Use the formula

$$\int_0^\infty t^{\alpha-1} \cos st \, dt = \Gamma(\alpha) |s|^{-\alpha} \cos\left(\tfrac{1}{2}\alpha\pi\right) , \qquad 0 < \alpha < 1 \ ,$$

and show that the integral equation

$$f(s) = \int_0^1 [g(t)/|s-t|^\alpha]\, dt$$

has no more than one solution.

8. Solve the integral equation

$$\sum_{r=0}^{n} a_r s^r = \frac{1}{\pi} \int_0^{*\ell} \frac{g(t)\, dt}{t-s}\,,$$

where a_r are given constants.

   Hint: Make the substitutions

$$s = (\ell/2)(1-\cos\theta)\,, \qquad t = (\ell/2)(1-\cos\varphi)\,.$$

9. Use the method of Section 9.4 and find the resolvent for the integral equation

$$g(s) = f(s) + \int_0^s (s^2 - t^2) g(t)\, dt\,.$$

10. Solve the integral equation

$$g(s) = f(s) + \lambda \int_0^s J_1(s-t) g(t)\, dt\,.$$

11. Find the resolvent of the integral equation

$$g(s) = f(s) + \int_0^s \frac{(s-t)^{m-1}}{(m-1)!}\, g(t)\, dt\,,$$

and complete the solution of the Example 3 in Section 5.3.

12. Use the infinite Hilbert transform pair and solve the integral equation

$$1/(1+s^2) = \int_{-\infty}^{*\infty} [g(t)/(s-t)]\, dt\,.$$

APPLICATIONS TO MIXED
BOUNDARY VALUE PROBLEMS

Mixed boundary value problems occur in physical sciences rather frequently and various mathematical techniques have been used to solve them. In this chapter, we present an integral-equation method applicable to most of these problems.

10.1. TWO-PART BOUNDARY VALUE PROBLEMS

An integral equation of the form

$$\int_0^a K_0(t, \rho) g(t) \, dt = f(\rho), \qquad 0 < \rho < a, \qquad (1)$$

where the function $f(\rho)$ and the kernel $K_0(t, \rho)$ are known and $g(t)$ is to be evaluated, embodies the solution of various mixed boundary value problems in potential theory, elastostatics, steady heat conduction, the flow of perfect fluids, and various other problems of equilibrium states. The boundaries involved are those of solids such as circular disks, elliptic disks, spherical caps, and spheroidal caps.

The integral equation (1) is a Fredholm integral equation of the first kind and is therefore, in general, difficult to solve. However, it is possible to reduce the solution of (1) to that of a pair of Volterra integral equations of the first kind with rather simple kernels. This reduction is achieved for every kernel $K_0(t, \rho)$ that for all $g(t)$ satisfies the relation

$$\int_0^a K_0(t,\rho)g(t)\,dt = h_1(\rho) \int_0^\rho K_2(w,\rho)\,[h_2(w)]^2$$

$$\times \int_w^a K_2(w,t)g(t)h_3(t)\,dt\,dw, \qquad 0 < \rho < a, \quad (2)$$

where h_1, h_2, h_3, and K_2 are known functions. It is further assumed that the kernel K_2 is such that the Volterra integral equations

$$\int_0^\rho K_2(t,\rho)g(t)\,dt = f(\rho), \qquad 0 < \rho < a, \tag{3}$$

and

$$\int_\rho^a K_2(\rho,t)g(t)\,dt = f(\rho), \qquad 0 < \rho < a \tag{4}$$

possess explicit unique solutions for g in terms of f, for all arbitrary differentiable functions f.

The task of solving (1) is now readily accomplished if we define two functions $S(\rho)$ and $C(\rho)$ such that

$$S(\rho) = h_2(\rho) \int_\rho^a K_2(\rho,t)g(t)h_3(t)\,dt, \qquad 0 < \rho < a, \tag{5}$$

and

$$f(\rho) = h_1(\rho) \int_0^\rho K_2(w,\rho)C(w)h_2(w)\,dw, \qquad 0 < \rho < a. \tag{6}$$

With the help of relations (2), (5), and (6), equation (1) takes the form

$$h_1(\rho) \int_0^\rho K_2(w,\rho)h_2(w)S(w)\,dw = h_1(\rho) \int_0^\rho K_2(w,\rho)C(w)h_2(w)\,dw, \tag{7}$$

$$0 < \rho < a \, ,$$

or

$$S(\rho) = C(\rho) \, , \qquad 0 < \rho < a \, . \tag{8}$$

In view of the assumptions already made about the solutions of the Volterra integral equations (3) and (4), we can solve equation (6) for the function $C(\rho)$ and hence, from (8), $S(\rho)$ is known. We can then invert the integral equation (5) and obtain the required function $g(t)$. We illustrate the above analysis with the following example.

Example.   Solve the integral equation

$$\int_0^a t\phi(t) \int_0^\infty J_1(p\rho)J_1(pt) \, dp \, dt = \Omega\rho \, , \qquad 0 < \rho < a \, , \tag{9}$$

where $\phi(t)$ is the unknown function. This equation solves the problem of the torsion of an isotropic and homogeneous elastic half-space due to a uniformly rotating, rigid circular disk which is attached to its free face (see Section 6.6, Example 2). The function $J_1(x)$ is the Bessel function. Comparing (1) and (9), we have

$$g(t) = t\phi(t) \, , \qquad f(\rho) = \Omega\rho \, , \qquad K_0(t,\rho) = \int_0^\infty J_1(p\rho)J_1(pt) \, dp \, . \tag{10}$$

The kernel K_0 satisfies the relation (2) because, for all $g(t)$, we can write

$$\int_0^a K_0(t,\rho)g(t) \, dt = \int_0^a g(t) \int_0^\infty J_1(p\rho)J_1(pt) \, dp \, dt$$

$$= \int_0^a g(t) \int_0^\infty \frac{2p}{\pi\rho t} \int_0^\rho \int_0^t \frac{J_{1/2}(pw)J_{1/2}(pv)(wv)^{3/2} \, dv \, dw \, dp \, dt}{(\rho^2 - w^2)^{1/2}(t^2 - v^2)^{1/2}}$$

$$= \frac{2}{\pi\rho} \int_0^a t^{-1} g(t) \int_0^\rho \int_0^t \frac{\delta(w-v)(wv) \, dv \, dw \, dt}{(\rho^2 - w^2)^{1/2}(t^2 - v^2)^{1/2}}$$

$$= \frac{2}{\pi\rho} \int\limits_0^a t^{-1} g(t) \int\limits_0^{\min(\rho,t)} \frac{w^2 \, dw \, dt}{(\rho^2 - w^2)^{\frac{1}{2}} (t^2 - w^2)^{\frac{1}{2}}}$$

$$= \frac{2}{\pi\rho} \int\limits_0^\rho \frac{w^2}{(\rho^2 - w^2)^{\frac{1}{2}}} \int\limits_w^a \frac{t^{-1} g(t) \, dt \, dw}{(t^2 - w^2)^{\frac{1}{2}}} , \qquad 0 < \rho < a , \qquad (11)$$

where we have used the first Sonine integral:

$$J_n(p\rho) = \left(\frac{2p}{\pi}\right)^{\frac{1}{2}} \frac{1}{\rho^n} \int\limits_0^\rho \frac{J_{n-(\frac{1}{2})}(pw) w^{n+(\frac{1}{2})}}{(\rho^2 - w^2)^{\frac{1}{2}}} \, dw , \qquad (12)$$

and the relation

$$\int\limits_0^\infty p J_\mu(pw) J_\mu(pv) \, dp = \delta(w-v)/(wv)^{\frac{1}{2}} , \qquad (13)$$

with δ the Dirac delta function. We have further used the sifting property of this function and changed the order of integration as explained in Figure 10.1.

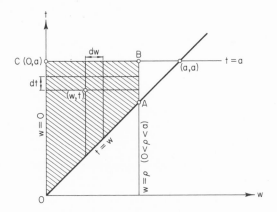

Figure 10.1

Comparing (2) and (11), we obtain the values of the functions h_1, h_2, h_3, and K_2 as

$$h_1(\rho) = 2/\pi\rho, \qquad h_2(\rho) = \rho, \qquad h_3(\rho) = 1/\rho,$$

$$K_2(t, \rho) = (\rho^2 - t^2)^{-\frac{1}{2}}. \tag{14}$$

Furthermore, the kernel K_2 is simple enough to ensure the inversion of the integral equations (3) and (4) (see Example 3 of Section 8.2). The present method is therefore applicable. Indeed, we set

$$S(\rho) = \rho \int_{\rho}^{a} \frac{\phi(t)\, dt}{(t^2 - \rho^2)^{\frac{1}{2}}}, \tag{15}$$

$$\Omega\rho = \frac{2}{\pi\rho} \int_{0}^{\rho} \frac{w S(w)\, dw}{(\rho^2 - w^2)^{\frac{1}{2}}}, \tag{16}$$

and the integral equation (9) is identically satisfied.

Finally, we invert (16) and obtain

$$S(\rho) = \frac{\Omega}{\rho} \frac{d}{d\rho} \int_{0}^{\rho} \frac{t^3\, dt}{(\rho^2 - t^2)^{\frac{1}{2}}} = 2\Omega\rho, \tag{17}$$

and then (15) yields the value of the function $\phi(\rho)$ as

$$\phi(\rho) = -\frac{4\Omega}{\pi} \frac{d}{d\rho} \int_{\rho}^{a} \frac{u\, du}{(u^2 - \rho^2)^{\frac{1}{2}}} = \frac{4\Omega\rho}{\pi(a^2 - \rho^2)^{\frac{1}{2}}}. \tag{18}$$

10.2. THREE-PART BOUNDARY VALUE PROBLEMS

A three-part boundary value problem has an integral representation formula of the form

$$\int_b^a K_0(t,\rho)g(t)\,dt = f(\rho)\,, \qquad b < \rho < a\,, \tag{1}$$

where b and a are two given numbers such as the inner and outer radii of an annular disk or the bounding angles of an annular spherical cap. The function f and the kernel K_0 are known, while g is to be determined. Let us set

$$f(\rho) = \sum_{r=-\infty}^{\infty} a_r \rho^r = f_1(\rho) + f_2(\rho)\,, \tag{2}$$

where

$$f_1(\rho) = \sum_{r=0}^{\infty} a_r \rho^r\,, \qquad 0 \leqslant \rho < a\,, \tag{3}$$

and

$$f_2(\rho) = \sum_{r=-\infty}^{-1} a_r \rho^r\,, \qquad b < \rho < \infty\,. \tag{4}$$

In addition, we define two functions $g_1(\rho)$ and $g_2(\rho)$ such that

$$g_1(\rho) + g_2(\rho) = \begin{cases} 0\,, & 0 \leqslant \rho < b\,, \\ g(\rho)\,, & b \leqslant \rho \leqslant a\,, \\ 0\,, & a < \rho < \infty\,. \end{cases} \tag{5}$$

From the relations (2)–(5), it follows that the integral equation (1) splits into two integral equations:

$$\int_0^\infty K_0(t,\rho)g_1(t)\,dt = f_1(\rho)\,, \qquad 0 < \rho < a\,, \tag{6}$$

and

$$\int_0^\infty K_0(t,\rho)g_2(t)\,dt = f_2(\rho)\,, \qquad b < \rho < \infty\,. \tag{7}$$

Proceeding as in Section 2, we assume that the kernel $K_0(t,\rho)$ is such that for all $g(t)$ it satisfies the relation

$$\int_0^\infty K_0(t,\rho) g(t)\, dt = \begin{cases} h_{11}(\rho) \int_0^\rho K_2(w,\rho) [h_{12}(w)]^2 \\[2mm] \quad \times \int_w^\infty K_2(w,t) g(t) h_{13}(t)\, dt\, dw\,, \qquad 0 < \rho < \infty, \\[4mm] h_{21}(\rho) \int_\rho^\infty K_2(\rho,w) [h_{22}(w)]^2 \\[2mm] \quad \times \int_0^w K_2(t,w) g(t) h_{23}(t)\, dt\, dw\,, \qquad 0 < \rho < \infty, \end{cases} \tag{8}$$

where h_{ij} $(i = 1, 2; j = 1, 2, 3)$, and the kernel K_2 are known functions. Moreover, the kernel K_2 is such that the Volterra integral equations

$$\int_0^\rho K_2(t,\rho) g(t)\, dt = f(\rho)\,, \qquad 0 < \rho < \infty\,, \tag{9}$$

and

$$\int_\rho^\infty K_2(\rho,t) g(t)\, dt = f(\rho)\,, \qquad 0 < \rho < \infty\,, \tag{10}$$

possess unique solutions for g in terms of all arbitrary differentiable functions f.

From relation (6) and the first part of relation (8), we have

$$h_{11}(\rho) \int_0^\rho K_2(w,\rho) [h_{12}(w)]^2 \int_w^\infty K_2(w,t) g_1(t) h_{13}(t)\, dt\, dw = f_1(\rho)\,, \tag{11}$$

$$0 < \rho < a\,.$$

Similarly, relation (7) and the second part of (8) give

$$h_{21}(\rho) \int_\rho^\infty K_2(\rho,w) [h_{22}(w)]^2 \int_0^w K_2(t,w) g_2(t) h_{23}(t)\, dt\, dw = f_2(\rho)\,, \tag{12}$$

$$b < \rho < \infty\,.$$

The next step is to define unknown functions $S_1, S_2, T_1, T_2, C_1,$ and C_2 such that

$$h_{12}(\rho) \int_\rho^\infty K_2(\rho, t) g_1(t) h_{13}(t)\, dt = \begin{cases} S_1(\rho), & 0 < \rho < a, \\ -T_1(\rho), & a < \rho < \infty, \end{cases} \quad (13)$$

$$h_{22}(\rho) \int_0^\rho K_2(t, \rho) g_2(t) h_{23}(t)\, dt = \begin{cases} -T_2(\rho), & 0 < \rho < b, \\ S_2(\rho), & b < \rho < \infty, \end{cases} \quad (14)$$

$$h_{11}(\rho) \int_0^\rho K_2(w, \rho) C_1(w) h_{12}(w)\, dw = f_1(\rho), \qquad 0 < \rho < a, \quad (15)$$

$$h_{21}(\rho) \int_\rho^\infty K_2(\rho, w) C_2(w) h_{22}(w)\, dw = f_2(\rho), \qquad b < \rho < \infty. \quad (16)$$

These Volterra-type integral equations are similar to equations (9) and (10), whose solutions are assumed to be known. From (11), (13), and (15), we derive the equation

$$h_{11}(\rho) \int_0^\rho K_2(w, \rho) h_{12}(w) S_1(w)\, dw$$

$$= h_{11}(\rho) \int_0^\rho K_2(w, \rho) C_1(w) h_{12}(w)\, dw, \qquad 0 < \rho < a, \quad (17)$$

or

$$S_1(\rho) = C_1(\rho), \qquad 0 < \rho < a. \quad (18)$$

Similarly, the result of combining (12), (14), and (16) is

$$S_2(\rho) = C_2(\rho), \qquad b < \rho < \infty. \quad (19)$$

The functions C_1 and C_2 can be evaluated in terms of the known functions f_1 and f_2 from equations (15) and (16). Hence, S_1 and S_2 are known. That leaves two unknown functions T_1 and T_2 still to be evaluated. For this purpose, we appeal to relations (5) and (13). The result is

$$h_{12}(\rho) \int_\rho^\infty K_2(\rho, t) g_2(t) h_{13}(t)\, dt = T_1(\rho), \qquad a < \rho < \infty. \quad (20)$$

Similarly, from (5) and (14), we have

$$h_{22}(\rho) \int_0^\rho K_2(t,\rho) g_1(t) h_{23}(t)\, dt = T_2(\rho)\,, \qquad 0 < \rho < b\,. \qquad (21)$$

Now, invert (14) to find the value of $g_2(t)$ in terms of T_2 and S_2 and substitute this value of $g_2(t)$ in (20). There results an integral equation containing the unknown functions T_1 and T_2. Likewise, the relations (13) and (21) lead to a second integral equation for T_1 and T_2. Both these equations are Fredholm integral equations of the second kind, and can therefore be solved by a straightforward iterative method.

Example.   Solve the integral equation

$$\int_b^a t\phi(t) \int_0^\infty J_1(p\rho) J_1(pt)\, dp\, dt = \Omega\rho\,, \qquad b < \rho < a\,, \qquad (22)$$

which embodies the solution of the torsion of an isotropic and homogeneous elastic half-space due to a uniformly rotating annular disk with inner radius b and outer radius a. Comparing it with (1) and (5), we have

$$g(t = t\phi(t)\,, \qquad g_1(t) = t\phi_1(t)\,, \qquad g_2(t) = t\phi_2(t)\,, \qquad (23)$$

$$f(\rho) = \Omega\rho\,, \qquad f_1(\rho) = \Omega\rho\,, \qquad 0 \leqslant \rho < a\,; \qquad f_2(\rho) = 0\,, \qquad (24)$$

$$b < \rho < \infty\,;$$

$$K_0(t,\rho) = \int_0^\infty J_1(p\rho) J_1(pt)\, dp\,, \qquad (25)$$

where

$$\phi_1(\rho) + \phi_2(\rho) = \begin{cases} 0\,, & 0 \leqslant \rho < b\,, \\ \phi(\rho)\,, & b \leqslant \rho \leqslant a\,, \\ 0\,, & a < \rho < \infty\,. \end{cases} \qquad (26)$$

In addition, the kernel $K_0(t,\rho)$ satisfies the requirement (8) in as much as, for all $g(t)$, we have

$$\int_0^\infty K_0(t,\rho) g(t)\, dt = \int_0^\infty g(t) \int_0^\infty J_1(p\rho) J_1(pt)\, dp\, dt$$

$$= \int_0^\infty g(t) \int_0^\infty \frac{2p}{\pi \rho t} \int_0^\rho \int_0^t \frac{J_{1/2}(pw) J_{1/2}(pv) (wv)^{3/2} \, dv \, dw \, dp \, dt}{(\rho^2 - w^2)^{1/2} (t^2 - v^2)^{1/2}}$$

$$= \frac{2}{\pi \rho} \int_0^\infty t^{-1} g(t) \int_0^\rho \int_0^t \frac{\delta(w - v)(wv) \, dv \, dw \, dt}{(\rho^2 - w^2)^{1/2}(t^2 - v^2)^{1/2}}$$

$$= \frac{2}{\pi \rho} \int_0^\infty t^{-1} g(t) \int_0^{\min(\rho,t)} \frac{w^2 \, dw \, dt}{(\rho^2 - w^2)^{1/2}(t^2 - w^2)^{1/2}}$$

$$= \frac{2}{\pi \rho} \int_0^\rho \frac{w^2}{(\rho^2 - w^2)^{1/2}} \int_w^\infty \frac{t^{-1} g(t) \, dt \, dw}{(t^2 - w^2)^{1/2}}, \qquad 0 < \rho < \infty, \qquad (27)$$

and

$$\int_0^\infty K_0(t, \rho) g(t) \, dt = \int_0^\infty g(t) \int_0^\infty J_1(p\rho) J_1(pt) \, dp \, dt$$

$$= \int_0^\infty g(t) \int_0^\infty \frac{2p\rho t}{\pi} \int_\rho^\infty \int_t^\infty \frac{J_{3/2}(pw) J_{3/2}(pv) \, dv \, dw \, dp \, dt}{(wv)^{1/2} (w^2 - \rho^2)^{1/2} (v^2 - t^2)^{1/2}}$$

$$= \frac{2\rho}{\pi} \int_0^\infty t g(t) \int_\rho^\infty \int_t^\infty \frac{\delta(w - v) \, dv \, dw \, dt}{(wv)(w^2 - \rho^2)^{1/2}(v^2 - t^2)^{1/2}}$$

$$= \frac{2\rho}{\pi} \int_0^\infty t g(t) \int_{\max(\rho,t)}^\infty \frac{w^{-2} \, dw \, dt}{(w^2 - \rho^2)^{1/2}(w^2 - t^2)^{1/2}}$$

$$= \frac{2\rho}{\pi} \int_\rho^\infty \frac{w^{-2}}{(w^2 - \rho^2)^{1/2}} \int_0^w \frac{t g(t) \, dt \, dw}{(w^2 - t^2)^{1/2}}, \qquad 0 < \rho < \infty, \qquad (28)$$

where we have used the formulas (10.1.12) and (10.1.13) and the equation

$$J_n(p\rho) = \left(\frac{2p}{\pi}\right)^{1/2} \rho^n \int_\rho^\infty \frac{J_{n+(1/2)}(pw) \, w^{-[n-(1/2)]}}{(w^2 - \rho^2)^{1/2}} \, dw \, .$$

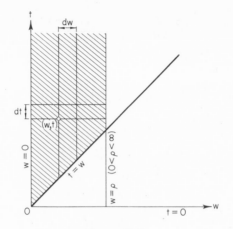

Figure 10.2

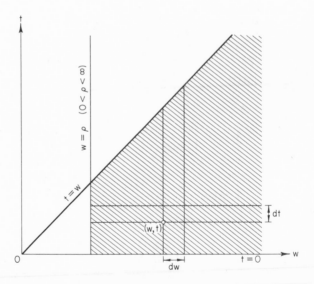

Figure 10.3

Furthermore, we have changed the order of integration in the steps leading to formulas (27) and (28) as explained in Figures 10.2 and 10.3. Hence,

$$h_{11}(\rho) = 2/\pi\rho, \qquad h_{12}(\rho) = \rho, \qquad h_{13}(\rho) = 1/\rho,$$

$$h_{21}(\rho) = 2\rho/\pi, \qquad h_{22}(\rho) = 1/\rho, \qquad h_{23}(\rho) = \rho, \qquad (29)$$

$$K_2(t,\rho) = (\rho^2 - t^2)^{-\frac{1}{2}}.$$

Furthermore, the kernel K_2 is such that the Volterra integral equations (9) and (10) can be readily solved, and therefore the method of this section can be applied.

The system of integral equations that corresponds to the system (13)–(21) is

$$\rho \int_\rho^\infty \frac{\phi_1(t)\, dt}{(t^2 - \rho^2)^{\frac{1}{2}}} = \begin{cases} S_1(\rho), & 0 < \rho < a, \\ -T_1(\rho), & a < \rho < \infty, \end{cases} \qquad (30)$$

$$\rho^{-1} \int_0^\rho \frac{t^2\, \phi_2(t)\, dt}{(\rho^2 - t^2)^{\frac{1}{2}}} = \begin{cases} -T_2(\rho), & 0 < \rho < b, \\ S_2(\rho), & b < \rho < \infty, \end{cases} \qquad (31)$$

$$\frac{2}{\pi\rho} \int_0^\rho \frac{w S_1(w)\, dw}{(\rho^2 - w^2)^{\frac{1}{2}}} = \Omega\rho, \qquad 0 < \rho < a, \qquad (32)$$

$$\frac{2\rho}{\pi} \int_\rho^\infty \frac{S_2(w)\, dw}{w(w^2 - \rho^2)^{\frac{1}{2}}} = 0, \qquad b < \rho < \infty, \qquad (33)$$

$$\rho \int_\rho^\infty \frac{\phi_2(t)\, dt}{(t^2 - \rho^2)^{\frac{1}{2}}} = T_1(\rho), \qquad a < \rho < \infty, \qquad (34)$$

$$\rho^{-1} \int_0^\rho \frac{t^2\, \phi_1(t)\, dt}{(\rho^2 - t^2)^{\frac{1}{2}}} = T_2(\rho), \qquad 0 < \rho < b. \qquad (35)$$

The integral equations (30)–(33) are readily inverted and the results are (see Example 3 of Section 8.2 and Example 4 of Section 9.5)

$$\phi_1(\rho) = -\frac{2}{\pi} \frac{d}{d\rho} \left[\int_\rho^a \frac{S_1(u)\, du}{(u^2 - \rho^2)^{\frac{1}{2}}} - \int_a^\infty \frac{T_1(u)\, du}{(u^2 - \rho^2)^{\frac{1}{2}}} \right], \qquad (36)$$

$$\phi_2(\rho) = \frac{2}{\pi\rho^2} \frac{d}{d\rho} \left[-\int_0^b \frac{u^2 \, T_2(u) \, du}{(\rho^2 - u^2)^{\frac{1}{2}}} + \int_b^\rho \frac{u^2 \, S_2(u) \, du}{(\rho^2 - u^2)^{\frac{1}{2}}} \right], \qquad (37)$$

$$S_1(\rho) = \frac{\Omega}{\rho} \frac{d}{d\rho} \int_0^\rho \frac{t^3 \, dt}{(\rho^2 - t^2)^{\frac{1}{2}}} = 2\Omega\rho, \qquad (38)$$

$$S_2(\rho) = 0. \qquad (39)$$

Substituting these values in (34) and (35), we get

$$T_1(\rho) = \frac{2\rho}{\pi} \int_0^b u^2 \, T_2(u) \int_\rho^\infty \frac{t^{-1} \, dt \, du}{(t^2 - \rho^2)^{\frac{1}{2}} (t^2 - u^2)^{\frac{3}{2}}}$$

$$= \frac{1}{\rho(\pi)^{\frac{1}{2}} \, \Gamma(5/2)} \int_0^b \frac{u^2 \, T_2(u) \, {}_2F_1(1/2, 1; 5/2; u^2/\rho^2) \, du}{(\rho^2 - u^2)}, \qquad (40)$$

$$a < \rho < \infty,$$

and

$$T_2(\rho) = \frac{4\Omega}{\pi\rho} \int_0^\rho \frac{t^3 \, dt}{(\rho^2 - t^2)^{\frac{1}{2}} (a^2 - t^2)^{\frac{1}{2}}} + \frac{2}{\pi\rho} \int_0^\rho \frac{t^3}{(\rho^2 - t^2)^{\frac{1}{2}}} \int_a^\infty \frac{T_1(u) \, du \, dt}{(u^2 - t^2)^{\frac{3}{2}}}$$

$$= \frac{8\Omega\rho^2 \, a^3}{3\pi(\rho^2 + a^2)^2} \, {}_2F_1\left(1, \frac{3}{2}; \frac{5}{2}; \frac{4a^2 \rho^2}{(\rho^2 + a^2)^2}\right)$$

$$+ \frac{\rho^2}{(\pi)^{\frac{1}{2}} \, \Gamma(5/2)} \int_a^\infty \frac{T_1(u) \, {}_2F_1(1/2, 1; 5/2; \rho^2/u^2) \, du}{u(u^2 - \rho^2)}, \qquad (41)$$

$$0 < \rho < b.$$

In the above relations, ${}_2F_1$ stands for the hypergeometric function and we have used the following relations pertaining to this function:

$$\int_\rho^\infty \frac{t^{-1} \, dt}{(t^2 - \rho^2)^{\frac{1}{2}} (t^2 - u^2)^{\frac{3}{2}}} = \frac{\pi^{\frac{1}{2}} \rho^{-2}}{2\Gamma(5/2)(\rho^2 - u^2)} \, {}_2F_1\left(\frac{1}{2}, 1; \frac{5}{2}; \frac{u^2}{\rho^2}\right),$$

$$u < \rho,$$

$$\int_0^\rho \frac{t^3\,dt}{(\rho^2-t^2)^{1/2}(u^2-t^2)^{3/2}} = \frac{\pi^{1/2}\rho^3\,{}_2F_1(1/2,1;5/2;\rho^2/u^2)}{2\Gamma(5/2)u(u^2-\rho^2)}\,, \qquad \rho < u\,,$$

$$\int_0^\rho \frac{t^3\,dt}{(\rho^2-t^2)^{1/2}(a^2-t^2)^{1/2}} = \frac{1}{2}\frac{\pi^{1/2}(a\rho)^3}{\Gamma(5/2)(\rho^2+a^2)^2}\,{}_2F_1\!\left(1,\frac{3}{2};\frac{5}{2};\frac{4a^2\rho^2}{(\rho^2+a^2)^2}\right),$$

$$\rho < a\,.$$

Equations (40) and (41) are two simultaneous Fredholm integral equations of the second kind and can be solved approximately by iteration when we introduce the parameter $\lambda = b/a$, such that $\lambda \ll 1$. Indeed, the hypergeometric function ${}_2F_1$ occurring under the integral signs in these equations is reducible to an elementary function:

$$_2F_1\!\left(\frac{1}{2},1;\frac{5}{2};\frac{x^2}{y^2}\right) = \frac{3y}{4x^3}\left[2xy - (y^2-x^2)\log\!\left(\frac{y+x}{y-x}\right)\right], \qquad x < y. \quad (42)$$

Thereby equations (40) and (41) take the simple forms

$$T_1(a\rho) = \frac{1}{\pi}\int_0^1 T_2(bu)\left[\frac{2\lambda\rho}{\rho^2-\lambda^2u^2} - \frac{1}{u}\log\!\left(\frac{\rho+\lambda u}{\rho-\lambda u}\right)\right]du\,, \qquad (43)$$

$$1 < \rho < \infty\,,$$

and

$$T_2(b\rho) = \frac{8\Omega\lambda^2\rho^2 a}{3\pi(1+\lambda^2\rho^2)^2}\,{}_2F_1\!\left(1,\frac{3}{2};\frac{5}{2};\frac{4\lambda^2\rho^2}{(1+\lambda^2\rho^2)^2}\right)$$

$$+ \frac{1}{\lambda\rho\pi}\int_1^\infty T_1(au)\left[\frac{2\lambda u\rho}{u^2-\lambda^2\rho^2} - \log\!\left(\frac{u+\lambda\rho}{u-\lambda\rho}\right)\right]du\,,$$

$$0 < \rho < 1\,. \qquad (44)$$

We first attend to (44) and observe that

$$_2F_1\!\left(1,\frac{3}{2};\frac{5}{2};x^2\right) = 1 + 3\sum_{n=1}^\infty \frac{x^{2n}}{2n+3}\,. \qquad (45)$$

With the help of this value, we readily obtain the first iteration for T_2 as

$$T_2(b\rho) = \frac{8\Omega a}{3\pi}\left\{\lambda^2\rho^2 + \frac{2\rho^4\lambda^4}{5} + O(\lambda^6)\right\}. \tag{46}$$

This value in turn helps us in solving (43) approximately

$$T_1(a\rho) = \frac{32\Omega a\lambda^5}{45\pi^2}\left[\frac{1}{\rho^3}\left(1 + \frac{2}{7}\lambda^2\right) + \frac{6\lambda^2}{7\rho^5} + O(\lambda^4)\right]. \tag{47}$$

In the above approximations, we have included only those terms that are needed to evaluate the torque experienced by the annulus up to $O(\lambda^9)$.

Finally, we substitute the values of S_1, S_2, T_1, and T_2 as given by the relations (38), (39), (46), and (47) in (36) and (37), and get

$$\phi_1(\rho) = \frac{4\Omega}{\pi}\left(\frac{\rho/a}{[1-(\rho^2/a^2)]^{1/2}} + \frac{16\lambda^5}{45\pi^2}\left\{\left(1 + \frac{2\lambda^2}{7}\right)\frac{a^3}{2\rho^3}\right.\right.$$

$$\times\left[-\frac{3a}{\rho}\sin^{-1}\frac{\rho}{a} + \left(1 - \frac{\rho^2}{a^2}\right)^{1/2} + 2\left(1 - \frac{\rho^2}{a^2}\right)^{-1/2}\right] + \frac{3\lambda^2}{28}\frac{a^5}{\rho^5}$$

$$\times\left[-15\frac{a}{\rho}\sin^{-1}\frac{\rho}{a} - 2\left(1 - \frac{\rho^2}{a^2}\right)^{3/2} + 9\left(1 - \frac{\rho^2}{a^2}\right)^{1/2}\right.$$

$$\left.\left.+ 8\left(1 - \frac{\rho^2}{a^2}\right)^{-1/2}\right]\right\} + O(\lambda^9)\right), \tag{48}$$

$$\phi_2(\rho) = -\frac{4\Omega}{3\pi^2}\left\{2\lambda\left[\frac{3\rho}{b}\sin^{-1}\frac{b}{\rho} - \left(1 - \frac{b^2}{\rho^2}\right)^{1/2} - 2\left(1 - \frac{b^2}{\rho^2}\right)^{-1/2}\right]\right.$$

$$+ \frac{\lambda^3}{5}\frac{\rho^2}{b^2}\left[\left(\frac{15\rho}{b}\right)\sin^{-1}\frac{b}{\rho} + 2\left(1 - \frac{b^2}{\rho^2}\right)^{3/2} - 9\left(1 - \frac{b^2}{\rho^2}\right)^{1/2}\right.$$

$$\left.\left.- 8\left(1 - \frac{b^2}{\rho^2}\right)^{-1/2}\right] + O(\lambda^5)\right\}. \tag{49}$$

Substituting these values in the relation $\phi(\rho) = \phi_1(\rho) + \phi_2(\rho)$, we obtain the desired solution of the integral equation (1).

10.3. GENERALIZED TWO-PART BOUNDARY VALUE PROBLEMS

An integral equation of a more general type such as

$$\int_0^a g(t) K_1(t,\rho) \, dt = f(\rho), \qquad 0 < \rho < a, \tag{1}$$

where the kernels K_1 can be perturbed on the kernel $K_0(t,\rho)$ of Section 10.1, can also be solved by the present method. This necessitates the splitting of the kernel K_1 as

$$K_1(t,\rho) = K_0(t,\rho) + G(t,\rho), \tag{2}$$

where the kernel $G(t,\rho)$ is in some sense smaller than K_0. From (1) and (2), it follows that

$$\int_0^a K_0(t,\rho) g(t) \, dt = f(\rho) - \int_0^a G(t,\rho) g(t) \, dt, \qquad 0 < \rho < a. \tag{3}$$

The kernel $K_0(t,\rho)$ satisfies the same requirements as those in Section 10.1. From (10.1.2) and (3), we have

$$h_1(\rho) \int_0^\rho K_2(w,\rho) \, [h_2(w)]^2 \int_w^a g(t) h_3(t) K_2(w,t) \, dt \, dw$$

$$= f(\rho) - \int_0^a G(t,\rho) g(t) \, dt, \qquad 0 < \rho < a. \tag{4}$$

Now we attempt to put the right side of this equation in the form of the left side as we did in equation (10.1.7). This is done by defining two functions $S(\rho)$ and $C(\rho)$ as in (10.1.5) and (10.1.6) and a new function $L(v,w)$ such that

$$G(t,\rho) = h_1(\rho) h_3(t) \int_0^\rho \int_0^t K_2(w,\rho) K_2(v,t) h_2(w) h_2(v) L(v,w) \, dv \, dw. \tag{5}$$

Thus, the integral on the right side of (4) takes the form

$$\int_0^a G(t,\rho) g(t) \, dt = \int_0^a g(t) h_1(\rho) h_3(t) \qquad \text{(equation continued)}$$

$$\times \int_0^\rho \int_0^t K_2(w,\rho)\,K_2(v,t)\,h_2(w)\,h_2(v)\,L(v,w)\,dv\,dw\,dt$$

$$= h_1(\rho) \int_0^\rho K_2(w,\rho)\,h_2(w) \int_0^a L(v,w)\,h_2(v)$$

$$\times \int_v^a K_2(v,t)\,g(t)\,h_3(t)\,dt\,dv\,dw\,, \qquad 0 < \rho < a, \quad (6)$$

where we have assumed that various orders of integration can be interchanged. When we substitute (10.1.5), (10.1.6), and (6) in equation (4), we get

$$h_1(\rho) \int_0^\rho K_2(w,\rho)\,h_2(w)\,S(w)\,dw = h_1(\rho) \int_0^\rho K_2(w,\rho)\,C(w)\,h_2(w)\,dw$$

$$- h_1(\rho) \int_0^\rho K_2(w,\rho)\,h_2(w)$$

$$\times \int_0^a L(v,w)\,S(v)\,dv\,dw\,, \qquad (7)$$

$$0 < \rho < a.$$

From this equation, it follows that

$$S(\rho) = C(\rho) - \int_0^a L(v,\rho)\,S(v)\,dv\,, \qquad 0 < \rho < a, \qquad (8)$$

which is a Fredholm integral equation of the second kind and can be solved for $S(\rho)$. The required function $g(t)$ is then obtained by inverting (10.1.5).

Example. Solve the integral equation

$$\int_0^a t\phi(t) \int_0^\infty [pJ_1(p\rho)J_1(pt)/\gamma]\,dp\,dt = \Omega\rho\,, \qquad 0 < \rho < a, \qquad (9)$$

where

$$\gamma = \begin{cases} -i(k^2-p^2)^{1/2}\,, & k \geqslant p\,, \\ (p^2-k^2)^{1/2}\,, & p \geqslant k\,. \end{cases} \qquad (10)$$

As explained in Section 6.6, Example 2, this equation solves the problem of the torsional oscillations of an isotropic and homogeneous elastic half-space due to a rigid circular disk of radius a which is performing simple harmonic oscillations.

Comparison of (9) and (1) gives

$$g(t) = t\phi(t), \qquad f(\rho) = \Omega\rho, \tag{11}$$

$$K_1(t,\rho) = \int_0^\infty [pJ_1(p\rho)J_1(pt)/\gamma]\, dp.$$

We split K_1 as in (2) with

$$K_0(t,\rho) = \int_0^\infty J_1(p\rho)J_1(pt)\, dp, \tag{12}$$

$$G(t,\rho) = \int_0^\infty \left(\frac{p}{\gamma} - 1\right) J_1(p\rho)J_1(pt)\, dp$$

$$= \frac{2}{\pi\rho t} \int_0^\rho \int_0^t \frac{(wv)^{3/2}}{(\rho^2 - w^2)^{1/2}(t^2 - w^2)^{1/2}}$$

$$\times \left[\int_0^\infty p \left(\frac{p}{\gamma} - 1\right) J_{1/2}(pv)J_{1/2}(pw)\, dp \right] dv\, dw. \tag{13}$$

By following the method of this section and borrowing the known results in Section 10.1, we have

$$h_1(\rho) = 2/\pi\rho, \qquad h_2(\rho) = \rho, \qquad h_3(\rho) = 1/\rho,$$

$$K_2(t,\rho) = (\rho^2 - t^2)^{-1/2}, \tag{14}$$

$$L(v,w) = (vw)^{1/2} \int_0^\infty [p(p/\gamma - 1)] J_{1/2}(pv)J_{1/2}(pw)\, dp, \tag{15}$$

$$S(\rho) = \rho \int_\rho^a \frac{\phi(t)\, dt}{(t^2 - \rho^2)^{1/2}}, \tag{16}$$

$$\Omega\rho = \frac{2}{\pi\rho} \int_0^\rho \frac{wC(w)\,dw}{(\rho^2 - w^2)^{\frac{1}{2}}}, \tag{17}$$

and

$$S(\rho) = C(\rho) - \int_0^a L(v, \rho)\, S(v)\, dv, \qquad 0 < \rho < a. \tag{18}$$

Equation (17) is readily inverted as shown in relation (10.1.17):

$$C(\rho) = 2\Omega\rho. \tag{19}$$

Therefore, equation (18) becomes

$$S(\rho) = 2\Omega\rho - \int_0^a L(v, \rho)\, S(v)\, dv, \qquad 0 < \rho < a. \tag{20}$$

The infinite integral (15) can be converted to a finite integral (see Appendix A.2):

$$L(v, w) = \begin{cases} i(vw)^{\frac{1}{2}} \int_0^k [p^2/(k^2 - p^2)^{\frac{1}{2}}]\, H_{\frac{1}{2}}^{(1)}(pv)\, J_{\frac{1}{2}}(pw)\, dp, \\[2mm] \qquad v \geqslant w, \\[4mm] i(vw)^{\frac{1}{2}} \int_0^k [p^2/(k^2 - p^2)^{\frac{1}{2}}]\, J_{\frac{1}{2}}(pv)\, H_{\frac{1}{2}}^{(1)}(pw)\, dp, \\[2mm] \qquad w \geqslant v, \end{cases} \tag{21}$$

where $H_{\frac{1}{2}}^{(1)}$ is a Hankel function of the first kind. This form of the kernel is useful for small values of k.

With this much information, we can solve the integral equation (20) approximately for small values of ak, which happens to be a dimensionless parameter. For this purpose, we write it as

$$S(a\rho) = 2\Omega a\rho - a \int_0^1 L(av, a\rho)\, S(av)\, dv, \qquad 0 < \rho < 1. \tag{22}$$

An approximate value of the kernel $aL(av, a\rho)$ is obtained from (21) by using the series expansions for the Hankel and Bessel functions. The result is

$$aL(av, a\rho) = \begin{cases} \dfrac{\alpha^2 \rho}{2} + \dfrac{4i\alpha^3 \rho v}{3\pi} - \dfrac{\alpha^4}{16}(3\rho v^2 + \rho^3) - \dfrac{8i\alpha^5}{45\pi}(\rho^3 v + \rho v^3) \\[2mm] \quad + \dfrac{\alpha^6}{384}(5\rho v^4 + 10\rho^3 v^2 + \rho^5) \\[2mm] \quad + \dfrac{4i\alpha^7}{1575\pi}(3\rho v^5 + 10\rho^3 v^3 + 3\rho^5 v) + O(\alpha^8), \\[2mm] \qquad v \geqslant \rho, \\[4mm] \dfrac{\alpha^2 v}{2} + \dfrac{4i\alpha^3 \rho v}{3\pi} - \dfrac{\alpha^4}{16}(3\rho^2 v + v^3) - \dfrac{8i\alpha^5}{45\pi}(\rho v^3 + \rho^3 v) \\[2mm] \quad + \dfrac{\alpha^6}{384}(5\rho^4 v + 10\rho^2 v^3 + v^5) \\[2mm] \quad + \dfrac{4i\alpha^7}{1575\pi}(3\rho v^5 + 10\rho^3 v^3 + 3\rho^5 v) + O(\alpha^8), \\[2mm] \qquad \rho \geqslant v, \end{cases} \tag{23}$$

where $x = ak$. By applying the straightforward iteration method to (22), we obtain an approximate value for $S(a\rho)$ as

$$S(a\rho) = 2\Omega a [c_1(\alpha)\rho + c_3(\alpha)\rho^3 + c_5(\alpha)\rho^5 + c_7(\alpha)\rho^7 + O(\alpha^8)], \tag{24}$$

where

$$c_1(\alpha) = 1 - \frac{\alpha^2}{4} - \frac{4i\alpha^3}{9\pi} + \frac{19\alpha^4}{192} + \frac{53i\alpha^5}{225\pi} - \left(\frac{16}{81\pi^2} + \frac{143}{3840}\right)\alpha^6 - \frac{8051i\alpha^7}{58800\pi},$$

$$c_3(\alpha) = \frac{\alpha^2}{12} + \frac{\alpha^4}{96} + \frac{i\alpha^5}{45\pi} - \frac{11\alpha^6}{2304} - \frac{47i\alpha^7}{4200\pi},$$

$$c_5(\alpha) = -\frac{\alpha^4}{960} - \frac{\alpha^6}{3840} - \frac{i\alpha^7}{1680\pi}, \qquad c_7(\alpha) = \frac{\alpha^6}{80640}.$$

Finally, we invert (16) to get

$$\phi(\rho) = -\frac{2}{\pi}\frac{d}{d\rho}\int_\rho^a \frac{S(u)\,du}{(u^2 - \rho^2)^{1/2}} = -\frac{2}{\pi}\frac{d}{d\rho}\int_{\rho/a}^1 \frac{S(au)\,du}{[u^2 - (\rho^2/a^2)]^{1/2}}. \tag{25}$$

From (24) and (25), we have

$$\phi(\rho) = \frac{4\Omega\rho/a}{\pi[1-(\rho^2/a^2)]^{1/2}} \left\{ 1 - \alpha^2 \left[\frac{1}{6} + \frac{1}{6} \left(1 - \frac{\rho^2}{a^2} \right) \right] - \frac{4i\alpha^3}{9\pi} \right.$$

$$+ \alpha^4 \left[\frac{13}{120} - \frac{1}{60} \left(1 - \frac{\rho^2}{a^2} \right) - \frac{1}{360} \left(1 - \frac{\rho^2}{a^2} \right)^2 \right]$$

$$+ \frac{i\alpha^5}{\pi} \left[\frac{58}{225} - \frac{2}{45} \left(1 - \frac{\rho^2}{a^2} \right) \right] - \alpha^6 \left[\left(\frac{16}{81\pi^2} + \frac{17}{1680} \right) \right.$$

$$\left. - \frac{53}{5040} \left(1 - \frac{\rho^2}{a^2} \right) + \frac{1}{1680} \left(1 - \frac{\rho^2}{a^2} \right)^2 + \frac{1}{25200} \left(1 - \frac{\rho^2}{a^2} \right)^3 \right]$$

$$\left. - \frac{i\alpha^7}{\pi} \left[\frac{1093}{7350} - \frac{13}{525} \left(1 - \frac{\rho^2}{a^2} \right) + \frac{1}{630} \left(1 - \frac{\rho^2}{a^2} \right)^2 \right] + O(\alpha^8) \right\}. \quad (26)$$

When $\alpha \to 0$, this reduces to equation (10.1.18).

10.4. GENERALIZED THREE-PART BOUNDARY VALUE PROBLEMS

Finally, we consider the integral equation

$$\int_b^a K_1(t,\rho) g(t) \, dt = f(\rho), \qquad b < \rho < a, \quad (1)$$

which is the generalization of the integral equation (10.2.1) and the kernel K_1 is to be perturbed on K_0 of Section 10.2. Indeed, we split it as

$$K_1(t,\rho) = K_0(t,\rho) + G(t,\rho), \quad (2)$$

and assume that $G(t,\rho)$ is in some sense smaller than K_0. With the help of the relations (10.2.2)–(10.2.5) and (2), the integral equation (1) becomes equivalent to the pair of equations

$$\int_0^\infty K_0(t,\rho) g_1(t) \, dt = f_1(\rho) - \int_0^\infty G(t,\rho) g_1(t) \, dt, \qquad 0 < \rho < a, \quad (3)$$

$$\int_0^\infty K_0(t,\rho)\,g_2(t)\,dt = f_2(\rho) - \int_0^\infty G(t,\rho)\,g_2(t)\,dt\,, \qquad b < \rho < \infty\,. \quad (4)$$

Furthermore, the choice of the kernel K_0 is such that the requirements embodied in the relations (10.2.8)–(10.2.10) are satisfied.

Now we extend the analysis of Section 10.3 and define two new kernels $L_1(v,w)$ and $L_2(v,w)$ such that

$$G(t,\rho) = \begin{cases} h_{11}(\rho)h_{13}(t)\displaystyle\int_0^\rho\int_0^t K_2(w,\rho)K_2(v,t)h_{12}(w)h_{12}(v) \\[2mm] \qquad\times L_1(v,w)\,dv\,dw\,, \\[4mm] h_{21}(\rho)h_{23}(t)\displaystyle\int_\rho^\infty\int_t^\infty K_2(\rho,w)K_2(t,v)h_{22}(w)h_{22}(v) \\[2mm] \qquad\times L_2(v,w)\,dv\,dw\,, \end{cases} \quad (5)$$

where the h's and the kernel K_2 are the same functions as occur in (10.2.8). Thus, the integrals on the right side of the relations (3) and (4) take the forms

$$\int_0^\infty G(t,\rho)\,g_1(t)\,dt = \int_0^\infty g_1(t)h_{11}(\rho)h_{13}(t)$$

$$\times \int_0^\rho\int_0^t K_2(w,\rho)K_2(v,t)h_{12}(w)h_{12}(v)$$

$$\times L_1(v,w)\,dv\,dw\,dt\,,$$

$$= h_{11}(\rho)\int_0^\rho K_2(w,\rho)h_{12}(w)\int_0^\infty L_1(v,w)h_{12}(v)$$

$$\times \int_v^\infty K_2(v,t)g_1(t)h_{13}(t)\,dt\,dv\,dw\,, \quad (6)$$

$$0 < \rho < a\,,$$

and

$$\int\limits_0^\infty G(t,\rho)g_2(t)\,dt = \int\limits_0^\infty g_2(t)h_{21}(\rho)h_{23}(t)$$

$$\times \int\limits_\rho^\infty \int\limits_t^\infty K_2(\rho,w)K_2(t,v)h_{22}(w)h_{22}(v)$$

$$\times L_2(v,w)\,dv\,dw\,dt$$

$$= h_{21}(\rho)\int\limits_\rho^\infty K_2(\rho,w)h_{22}(w)\int\limits_0^\infty L_2(v,w)h_{22}(v)$$

$$\times \int\limits_0^v K_2(t,v)g_2(t)h_{23}(t)\,dt\,dv\,dw, \tag{7}$$

$$b < \rho < \infty,$$

where we have assumed that various orders of integration may be interchanged.

From equations (3) and (6) and the first part of (10.2.8), we derive the relation

$$h_{11}(\rho)\int\limits_0^\rho K_2(w,\rho)[h_{12}(w)]^2\int\limits_w^\infty K_2(w,t)g_1(t)h_{13}(t)\,dt\,dw$$

$$= f_1(\rho) - h_{11}(\rho)\int\limits_0^\rho K_2(w,\rho)h_{12}(w)\int\limits_0^\infty L_1(v,w)h_{12}(v)$$

$$\times \int\limits_v^\infty K_2(v,t)g_1(t)h_{13}(t)\,dt\,dv\,dw, \qquad 0 < \rho < a. \tag{8}$$

Similarly, from equations (4) and (7) and the second part of (10.2.8), we have

$$h_{21}(\rho)\int\limits_\rho^\infty K_2(\rho,w)[h_{22}(w)]^2\int\limits_0^w K_2(t,w)g_2(t)h_{23}(t)\,dt\,dw$$

$$= f_2(\rho) - h_{21}(\rho)\int\limits_\rho^\infty K_2(\rho,w)h_{22}(w)\int\limits_0^\infty L_2(v,w)h_{22}(v)$$

$$\times \int\limits_0^v K_2(t,v)\, g_2(t)\, h_{23}(t)\, dt\, dv\, dw\,, \qquad b < \rho < \infty\,. \qquad (9)$$

The next step is to use in (8) and (9) the functions S_1, S_2, T_1, T_2, C_1, and C_2 as defined by the relations (10.2.13)–(10.2.16) and follow the arguments of Section 10.2. The functions C_1 and C_2 are known in terms of f_1 and f_2, while the four unknown functions S_1, S_2, T_1, and T_2 satisfy the following two simultaneous Fredholm integral equations of the second kind:

$$S_1(\rho) + \int\limits_0^a L_1(v,\rho)\, S_1(v)\, dv = C_1(\rho) + \int\limits_a^\infty L_1(v,\rho)\, T_1(v)\, dv\,, \qquad (10)$$

$$0 < \rho < a\,,$$

$$S_2(\rho) + \int\limits_b^\infty L_2(v,\rho)\, S_2(v)\, dv = C_2(\rho) + \int\limits_0^b L_2(v,\rho)\, T_2(v)\, dv\,, \qquad (11)$$

$$b < \rho < \infty\,.$$

The two Fredholm integral equations that are the results of equations (10.2.20) and (10.2.21) when the values of g_1 and g_2 are substituted in terms of S_1, T_1, S_2, and T_2 are the additional two equations that augment (10) and (11). Thereby, the system has become a determinate one and can be solved by iteration as in the previous sections.

Example. Solve the integral equation

$$\int\limits_b^a t\phi(t) \int\limits_0^\infty [pJ_1(p\rho)J_1(pt)/\gamma]\, dp\, dt = \Omega\rho\,, \qquad b < \rho < a\,, \qquad (12)$$

where γ is defined in (10.3.10). This equation governs the problem of torsional oscillations of an elastic half-space due to a rigid annular disk. The functions g, f, K_1, K_0, and G are the same as defined in (10.3.11)–(10.3.13). Similarly, the relations (10.2.27) and (10.2.28) remain valid in the present case, while the kernel $G(t,\rho)$ becomes

$$G(t,\rho) = \int\limits_0^\infty \left(\frac{p}{\gamma}-1\right) J_1(p\rho)J_1(pt)\, dp \qquad \begin{array}{r}(13)\\ \text{(continued)}\end{array}$$

$$
= \begin{cases}
\dfrac{2}{\pi \rho t} \displaystyle\int_0^\rho \int_0^t \dfrac{(wv)^{3/2}}{(\rho^2 - w^2)^{1/2}(t^2 - w^2)^{1/2}} \\[4mm]
\quad \times \left[\displaystyle\int_0^\infty p\left(\dfrac{p}{\gamma} - 1\right) J_{1/2}(pv) J_{1/2}(pw)\, dp \right] dv\, dw \,, \\[8mm]
\dfrac{2\rho t}{\pi} \displaystyle\int_\rho^\infty \int_t^\infty \dfrac{(wv)^{-1/2}}{(w^2 - \rho^2)^{1/2}(v^2 - t^2)^{1/2}} \\[4mm]
\quad \times \left[\displaystyle\int_0^\infty p\left(\dfrac{p}{\gamma} - 1\right) J_{3/2}(pv) J_{3/2}(pw)\, dp \right] dv\, dw \,,
\end{cases}
\tag{13}
$$

which corresponds to equation (5) above. Thus, the functions h_{ij} and the kernel K_2 are the same as defined by the relations (10.2.29), while the kernels L_1 and L_2 are

$$
L_1(v, w) = (wv)^{1/2} \int_0^\infty [p(p/\gamma - 1)] J_{1/2}(pv) J_{1/2}(pw)\, dp \,, \tag{14}
$$

$$
L_2(v, w) = (wv)^{1/2} \int_0^\infty [p(p/\gamma) - 1] J_{3/2}(pv) J_{3/2}(pw)\, dp \,. \tag{15}
$$

Note that $L_1(v, w)$ coincides with $L(v, w)$ as given by (10.3.15).

The four Fredholm integral equations of the second kind for the unknown functions S_1, S_2, T_1, and T_2 emerge as

$$
T_1(\rho) = \ell_2(\rho) + \frac{1}{\rho(\pi)^{1/2}\Gamma(5/2)}
$$

$$
\times \int_0^b \frac{u^2 T_2(u)\, {}_2F_1(1/2, 1; 5/2; u^2/\rho^2)\, du}{\rho^2 - u^2}, \qquad a < \rho < \infty, \tag{16}
$$

$$
T_2(\rho) = \ell_1(\rho) + \frac{\rho^2}{(\pi)^{1/2}\Gamma(5/2)}
$$

$$\times \int_a^\infty \frac{T_1(u)\,_2F_1(1/2,1;5/2;\rho^2/u^2)\,du}{u(u^2-\rho^2)}\,, \qquad 0 < \rho < b, \qquad (17)$$

$$S_1(\rho) + \int_0^a L_1(v,\rho)\,S_1(v)\,dv = 2\Omega\rho + \int_a^\infty L_1(v,\rho)\,T_1(v)\,dv\,, \qquad (18)$$

$$0 < \rho < a\,,$$

$$S_2(\rho) + \int_b^\infty L_2(v,\rho)\,S_2(v)\,dv = \int_0^b L_2(v,\rho)\,T_2(v)\,dv\,, \qquad (19)$$

$$b < \rho < \infty\,,$$

where

$$\ell_1(\rho) = -\frac{2}{\pi\rho}\int_0^\rho \frac{t^2}{(\rho^2-t^2)^{1/2}}\,\frac{d}{dt}\int_t^a \frac{S_1(u)\,du\,dt}{(u^2-t^2)^{1/2}}\,, \qquad 0 < \rho < b, \quad (20)$$

$$\ell_2(\rho) = \frac{2\rho}{\pi}\int_\rho^\infty \frac{1}{t^2(t^2-\rho^2)^{1/2}}\,\frac{d}{dt}\int_b^t \frac{u^2 S_2(u)\,du\,dt}{(t^2-u^2)^{1/2}}\,, \qquad (21)$$

$$a < \rho < \infty\,.$$

We solve equations (16)–(19) approximately by iteration when the parameters ka and b/a are small. In view of the relations (10.2.42), we can write equation (16) in the form

$$T_1(a\rho) = \ell_2(a\rho) + \frac{1}{\pi}\int_0^1 T_2(bu)\left[\frac{2\lambda\rho}{\rho^2-\lambda^2 u^2} - \frac{1}{u}\log\left(\frac{\rho+\lambda u}{\rho-\lambda u}\right)\right]du\,, \quad (22)$$

$$1 < \rho < \infty\,,$$

where $\lambda = b/a$. Similarly, (17) becomes

$$T_2(b\rho) = \ell_1(b\rho) + \frac{1}{\lambda\rho\pi}\int_1^\infty T_1(au)\left[\frac{2\lambda u\rho}{u^2-\lambda^2\rho^2} - \log\left(\frac{u+\lambda\rho}{u-\lambda\rho}\right)\right]du\,, \quad (23)$$

$$0 < \rho < 1\,.$$

Let us observe that the parameters that occur in this problem are

$$\alpha = ak, \qquad \beta = bk, \qquad \lambda = b/a = \beta/\alpha, \tag{24}$$

and the discussion in the sequel is based on the assumption that $\alpha = O(\lambda)$, and, as such, $\beta = \alpha\lambda = O(\alpha^2)$.

We start with equation (18) and write it as

$$S_1(a\rho) + a \int_0^1 L_1(av, a\rho) S_1(av) \, dv$$

$$= 2\Omega a\rho + a \int_1^\infty L_1(av, a\rho) T_1(av) \, dv, \qquad 0 < \rho < 1. \tag{25}$$

This is solved by setting

$$S_1(a\rho) = X_1(a\rho) + W_1(a\rho), \tag{26}$$

such that

$$X_1(a\rho) = 2\Omega a\rho - a \int_0^1 L_1(av, a\rho) X_1(av) \, dv, \qquad 0 < \rho < 1, \tag{27}$$

and

$$W_1(a\rho) = a \int_1^\infty L_1(av, a\rho) T_1(av) \, dv$$

$$- a \int_0^1 L_1(av, a\rho) W_1(av) \, dv, \qquad 0 < \rho < 1. \tag{28}$$

The integral equation (27) is precisely equation (10.3.22) since the kernels L and L_1 are identical. Therefore, $X_1(a\rho)$ is given by the expression on the right side of (10.3.24).

Similarly, the integral equation (19) can be written as

$$S_2(b\rho) + b \int_1^\infty L_2(bv, b\rho) S_2(bv) \, dv = b \int_0^1 L_2(bv, b\rho) T_2(bv) \, dv, \tag{29}$$

$$1 < \rho < \infty,$$

whose kernel can also be reduced to the following suitable form (see Appendix A.2).

$$L_2(v, \rho) = \begin{cases} i(\rho v)^{\frac{1}{2}} \int_0^k [p^2/(k^2 - p^2)^{\frac{1}{2}}] J_{\frac{3}{2}}(p\rho) H_{\frac{3}{2}}^{(1)}(pv) \, dp, \\[2mm] v \geqslant \rho, \\[4mm] i(\rho v)^{\frac{1}{2}} \int_0^k [p^2/(k^2 - p^2)^{\frac{1}{2}}] J_{\frac{3}{2}}(pv) H_{\frac{3}{2}}^{(1)}(p\rho) \, dp, \\[2mm] \rho \geqslant v. \end{cases} \tag{30}$$

Using the expansions for Bessel and Hankel functions, we readily derive the approximate formula

$$bL_2(bv, b\rho) = \begin{cases} \alpha^2 \lambda^2 [(\rho^2/6v) + O(\alpha^2)], & v \geqslant \rho, \\[2mm] \alpha^2 \lambda^2 [(v^2/6\rho) + O(\alpha^2)], & \rho \geqslant v. \end{cases} \tag{31}$$

The functions occurring in the system of equations (16)–(31) are to be calculated in the order $X_1, \ell_1, T_2, S_2, \ell_2, T_1, W_1, S_1$. Having found X_1, we can proceed to evaluate the other functions of this sequence. The required results, obtained by one iteration, are

$$\ell_1(b\rho) = \frac{8\Omega a \rho^2 \lambda^2}{3\pi} \left[\left(1 - \frac{\alpha^2}{3} - \frac{4i\alpha^3}{9\pi} \right) + \frac{2\rho^2 \lambda^2}{5} + O(\alpha^4) \right], \tag{32}$$

$$0 < \rho < 1,$$

$$T_2(b\rho) = \ell_1(b\rho) + O(\alpha^7), \qquad 0 < \rho < 1, \tag{33}$$

$$S_2(b\rho) = \frac{4\Omega a \alpha^2 \lambda^4}{45\pi} \left[\frac{1}{\rho} + O(\alpha^2) \right], \qquad 1 < \rho < \infty, \tag{34}$$

$$\ell_2(a\rho) = \frac{8\Omega a \alpha^2 \lambda^5}{45\pi^2} \left[\frac{1}{\rho} + O(\alpha^2) \right], \qquad 1 < \rho < \infty, \tag{35}$$

$$T_1(a\rho) = \frac{32\Omega a \lambda^5}{45\pi^2} \left[\frac{\alpha^2}{4\rho} + \frac{1}{\rho^3} \left(1 - \frac{\alpha^3}{3} + \frac{2}{7}\lambda^2 \right) + \frac{1}{\rho^5} \left(\frac{6\lambda^2}{7} \right) + O(\alpha^3) \right],$$

$$1 < \rho < \infty, \tag{36}$$

$$W_1(a\rho) = \frac{8\Omega a \alpha^2 \lambda^5}{45\pi^2} [\rho + O(\alpha)], \qquad 0 < \rho < 1, \tag{37}$$

$$S_1(a\rho) = X_1(a\rho) + \frac{8\Omega a\rho a^2 \lambda^5}{45\pi^2} + O(\alpha^8), \qquad 0 < \rho < 1. \tag{38}$$

In the above approximation, we have included only those terms that are needed to evaluate the value of the torque experienced by the annular disk to $O(\alpha^8)$.

The values of g_1 and g_2 are obtained by inverting (10.2.13) and (10.2.14) with S_1, S_2, T_1, and T_2 as given by the above formulas. Then, the value of the function g follows from (10.2.5) and that of $\phi(\rho)$ from the relation $\phi(\rho) = \rho^{-1} g(\rho)$.

Finally, let us note that the method explained in the previous sections requires the following modifications when applied to problems that relate to spherical caps and annular spherical caps (or spherical rings): change a to α, b to β, ρ to θ, ∞ to π, and replace the equations (10.2.2) by

$$f(\theta) = \sum_{r=-\infty}^{\infty} a_r \left(\tan\frac{\theta}{2} \right)^r = f_1(\theta) + f_2(\theta), \tag{39}$$

where

$$f_1(\theta) = \sum_{r=0}^{\infty} a_r \left(\tan\frac{\theta}{2} \right)^r, \qquad 0 \leqslant \theta < \alpha, \tag{40}$$

$$f_2(\theta) = \sum_{r=-\infty}^{-1} a_r \left(\tan\frac{\theta}{2} \right)^r, \qquad \beta < \theta < \pi. \tag{41}$$

Here, θ is the polar angle and α and β are the bounding angles of the annular cap.

10.5. FURTHER EXAMPLES

Example 1.   In Section 6.3, we found that the integral equation

$$f(\rho) = \int_0^a g(t) K_0(t, \rho)\, dt, \qquad 0 < \rho < a, \tag{1}$$

where

$$K_0(t,\rho) = 2\pi \int_0^\infty J_n(p\rho)J_n(pt)\,dp \tag{2}$$

embodies the electrostatic potential problem due to a circular disk of radius a charged to a prescribed potential $f(\rho)\cos n\varphi$. Following the analysis of Section 10.1, we get, for all $g(t)$,

$$\int_0^a K_0(t,\rho)g(t)\,dt = 2\pi \int_0^a g(t) \int_0^\infty J_n(p\rho)J_n(pt)\,dp\,dt$$

$$= 2\pi \int_0^a g(t) \int_0^\infty \frac{2p}{\pi(\rho t)^n}$$

$$\times \int_0^\rho \int_0^t \frac{J_{n-\frac12}(pw)J_{n-\frac12}(pv)(wv)^{n+\frac12}\,dv\,dw\,dp\,dt}{(\rho^2-w^2)^{\frac12}(t^2-v^2)^{\frac12}}$$

$$= \frac{4}{\rho^n} \int_0^a t^{-n}g(t) \int_0^\rho \int_0^t \frac{\delta(w-v)(wv)^n\,dv\,dw\,dt}{(\rho^2-w^2)^{\frac12}(t^2-v^2)^{\frac12}}$$

$$= \frac{4}{\rho^n} \int_0^a t^{-n}g(t) \int_0^{\min(\rho,t)} \frac{w^{2n}\,dw\,dt}{(\rho^2-w^2)^{\frac12}(t^2-w^2)^{\frac12}}$$

$$= \frac{4}{\rho^n} \int_0^\rho \frac{w^{2n}}{(\rho^2-w^2)^{\frac12}} \int_w^a \frac{t^{-n}g(t)\,dt\,dw}{(t^2-w^2)^{\frac12}}, \qquad 0 < \rho < a, \tag{3}$$

where we have used the relations (10.1.12) and (10.1.13).

Comparing (3) and (10.1.2), we obtain the values of the functions h_1, h_2, h_3, and K_2 as

$$h_1(\rho) = 4/\rho^n, \qquad h_2(\rho) = \rho^n, \qquad h_3(\rho) = \rho^{-n},$$
$$K_2(t,\rho) = (\rho^2-t^2)^{-\frac12}. \tag{4}$$

Moreover, this form of the kernel K_2 ensures the inversion of the integral equations (10.1.3) and (10.1.4). The method of Section 10.1 is therefore applicable in this case. Indeed, we have

$$S(\rho) = \rho^n \int_{\rho}^{a} \frac{t^{-n} g(t)\, dt}{(t^2 - \rho^2)^{1/2}}, \qquad 0 < \rho < a, \qquad (5)$$

$$f(\rho) = \frac{4}{\rho^n} \int_{0}^{\rho} \frac{w^n S(w)\, dw}{(\rho^2 - w^2)^{1/2}}, \qquad 0 < \rho < a. \qquad (6)$$

Inverting (6), we get

$$S(\rho) = \frac{\rho^{-n}}{2\pi} \frac{d}{d\rho} \int_{0}^{\rho} \frac{t^{n+1} f(t)\, dt}{(\rho^2 - t^2)^{1/2}}, \qquad (7)$$

which gives the value of the function S in terms of the known function f. Substituting this value of S in (5) and inverting it, we recover the value of the unknown function g:

$$g(t) = -\frac{2t^n}{\pi} \frac{d}{dt} \int_{t}^{a} \frac{w^{1-n} S(w)\, dw}{(w^2 - t^2)^{1/2}}. \qquad (8)$$

For this special case when the disk is kept at a unit potential, then $f(\rho) = 1$, $n = 0$. Equations (7) and (8) take the simple forms

$$S(\rho) = \frac{1}{2\pi} \frac{d}{d\rho} \int_{0}^{\rho} \frac{t\, dt}{(\rho^2 - t^2)^{1/2}} = \frac{1}{2\pi} \qquad (9)$$

and

$$g(t) = -\frac{1}{\pi^2} \frac{d}{dt} \int_{t}^{a} \frac{w\, dw}{(w^2 - t^2)^{1/2}} = \frac{t}{\pi^2 (a^2 - t^2)^{1/2}}, \qquad 0 < t < a. \quad (10)$$

Incidentally, we can evaluate the capacity C of the disk without finding the value of the unknown function g. Indeed, the formula for capacity is

$$C = 2\pi \int_{0}^{a} g(t)\, dt.$$

Substituting in it the value of $g(t)$ obtained from (8) after putting $n = 0$, we get

$$C = 4 \int_0^a S(w) \, dw = 2a/\pi \, . \tag{11}$$

Example 2.   The equation

$$1 = \int_0^a t g(t) K_1(t, \rho) \, dt \, , \qquad 0 < \rho < a \, , \tag{12}$$

where

$$K_1(t, \rho) = \int_0^\infty [p J_0(p\rho) J_0(pt)/\gamma] \, dp \, , \tag{13}$$

is the integral-equation formulation of the problem of acoustic diffraction of an axially symmetric plane wave by a perfectly soft circular disk of radius a (see Section 6.7, Example 1).

To solve (12), we split the kernel K_1 as

$$K_1(t, \rho) = K_0(t, \rho) + G(t, \rho) \, ,$$

where

$$K_0(t, \rho) = \int_0^\infty J_0(p\rho) J_0(pt) \, dp \tag{14}$$

and

$$G(t, \rho) = \int_0^\infty [(p/\gamma) - 1] J_0(p\rho) J_0(pt) \, dp \, . \tag{15}$$

Thereafter, the analysis is similar to the one given for the integral equation (10.3.9) of the example treated in Section 10.3.

Example 3.   *Electrostatic potential problem due to a spherical cap.* A spherical polar system (r, θ, ϕ) is chosen so that the cap is defined by $r = a$, $0 \leqslant \theta \leqslant \alpha$, $0 \leqslant \varphi \leqslant 2\pi$ (see Figure 10.4). We consider the axially symmetric case when the potential on the cap is given by $f(\theta)$. Thus the boundary value problem is

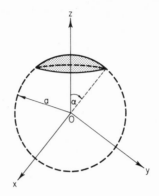

Figure 10.4

$$\nabla^2 V(r, \theta\, \varphi) = 0 \qquad \text{in} \quad D\,, \tag{16}$$

$$V(a, \theta, \varphi) = f(\theta)\,, \qquad 0 \leqslant \theta \leqslant \alpha\,; \quad 0 \leqslant \varphi \leqslant 2\pi\,; \tag{17}$$

where D is the region exterior to the cap. By following the method of Chapter 6 (see Exercise 9 of that chapter), we find that the integral representation formula for (16) is

$$V(r, \theta, \varphi) = a^2 \int_0^\alpha \int_0^{2\pi} [\sigma(t)/R] \sin t \, d\varphi_1 \, dt\,, \tag{18}$$

where $\sigma(t)$ is the charge density at the point $Q(a, t, \varphi_1)$ on the cap and

$$R = (r^2 + a^2 - 2ar \cos\gamma)^{1/2}, \quad \cos\gamma = \cos\theta\cos t + (\sin\theta\sin t)\cos(\varphi - \varphi_1).$$

Applying the boundary condition (17), we obtain the Fredholm integral equation of the first kind

$$f(\theta) = a^2 \int_0^\alpha (\sin t)\,\sigma(t)\, K_0(t, \theta)\, dt\,, \qquad 0 < \theta < \alpha\,, \tag{19}$$

where

$$K_0(t, \theta) = \int_0^{2\pi} \frac{d\varphi_1}{(2a^2 - 2a^2 \cos\gamma)^{1/2}}\,. \tag{20}$$

The next step is to expand the integrand in (20) in terms of the spherical harmonics $Y_n{}^m(\theta, \varphi)$:

$$\frac{1}{(2a^2 - 2a^2 \cos \gamma)^{\frac{1}{2}}} = \frac{1}{a} \sum_{n=0}^{n} \sum_{m=-n}^{n} \frac{(n-|m|)!\, Y_n{}^m(\theta, \varphi)\, Y_n^{m*}(t, \phi_1)}{(n+|m|)!} . \quad (21)$$

From (20) and (21), it follows that

$$K_0(t, \theta) = \frac{2\pi}{a} \sum_{n=0}^{\infty} P_n(\cos \theta) P_n(\cos t) , \quad (22)$$

where P_n is the Legendre polynomial. We can put (22) back into the integral form if we use the Mehler–Dirichlet integral

$$P_n(\cos \theta) = \frac{\sqrt{2}}{\pi} \int_0^{\theta} \frac{\cos[(n+\frac{1}{2})w]\, dw}{(\cos w - \cos \theta)^{\frac{1}{2}}} , \quad (23)$$

and the result

$$\sum_{n=0}^{\infty} \cos\left(n+\frac{1}{2}\right) w \cos\left(n+\frac{1}{2}\right) v = \frac{\pi}{2} \delta(w-v), \qquad 0 < w, v < \pi. \quad (24)$$

The kernel $K_0(t, \theta)$ as given by (22) then becomes

$$K_0(t, \theta) = \frac{2}{a} \int_0^{\theta} \int_0^{t} \frac{\delta(w-v)\, dv\, dw}{(\cos w - \cos \theta)^{\frac{1}{2}} (\cos v - \cos t)^{\frac{1}{2}}} \quad (25)$$

The relation that corresponds to (10.1.2) for the present case is

$$\int_0^{\alpha} K_0(t, \theta) g(t)\, dt = \frac{2}{a} \int_0^{\alpha} g(t) \int_0^{\min(\theta, t)} \frac{dw\, dt}{(\cos w - \cos \theta)^{\frac{1}{2}} (\cos w - \cos t)^{\frac{1}{2}}}$$

$$= \frac{2}{a} \int_0^{\theta} \frac{1}{(\cos w - \cos \theta)^{\frac{1}{2}}} \int_w^{\alpha} \frac{g(t)\, dt\, dw}{(\cos w - \cos t)^{\frac{1}{2}}} , \quad (26)$$

$$0 < \theta < \alpha .$$

Hence,

$$h_1(\theta) = 2/a, \qquad h_2(\theta) = h_3(\theta) = 1,$$

$$K_2(t, \theta) = (\cos t - \cos \theta)^{-\frac{1}{2}}. \tag{27}$$

Similarly, the relations that correspond to (10.1.5) and (10.1.6) are

$$S(\theta) = \int_\theta^\alpha \frac{a^2 \sigma(t) \sin t \, dt}{(\cos \theta - \cos t)^{\frac{1}{2}}}, \tag{28}$$

and

$$f(\theta) = \frac{2}{a} \int_0^\theta \frac{S(w) \, dw}{(\cos w - \cos \theta)^{\frac{1}{2}}}. \tag{29}$$

The integral equations (28) and (29) are simple Volterra integral equations and can be easily inverted. In fact, the inversion of (29) is readily achieved from Example 2 of Section 8.2, and we have

$$S(w) = \frac{a}{2\pi} \frac{d}{dw} \int_0^w \frac{f(\theta) \sin \theta \, d\theta}{(\cos \theta - \cos w)^{\frac{1}{2}}}. \tag{30}$$

Similarly, the solution of (28) is

$$\sigma(t) = -\frac{1}{\pi a^2 \sin t} \frac{d}{dt} \int_t^\alpha \frac{S(w) \sin w \, dw}{(\cos t - \cos w)^{\frac{1}{2}}}, \tag{31}$$

and the integral equation (19) is completely solved.

As pointed out in Example 1, one need not determine the charge density $\sigma(t)$ explicitly to find the capacity of the solid. Using the formula

$$C = 2\pi a^2 \int_0^\alpha (\sin t) \sigma(t) \, dt \tag{32}$$

and relation (31), it follows that

$$C = 2\pi a^2 \int_0^\alpha \left(-\frac{1}{\pi a^2}\right) \frac{d}{dt} \int_t^\alpha \frac{S(w) \sin w \, dw \, dt}{(\cos t - \cos w)^{\frac{1}{2}}}$$

$$= 2 \int_0^\alpha \frac{S(w)\sin w \, dw}{(1-\cos w)^{1/2}} = 2\sqrt{2} \int_0^\alpha S(w)\cos\left(\frac{w}{2}\right) dw \, . \tag{33}$$

For the special case when the cap is kept at a unit potential, that is, $f(\theta) = 1$, relation (30) simplifies to

$$S(w) = \frac{a}{2\pi} \frac{d}{dw} \int_0^w \frac{\sin\theta \, d\theta}{(\cos\theta - \cos w)^{1/2}} = \frac{a}{\sqrt{2\pi}} \cos\left(\frac{w}{2}\right) . \tag{34}$$

From (33) and (34), we have the value of the capacity as.

$$C = (a/\pi)(\alpha + \sin\alpha) \, . \tag{35}$$

EXERCISES

Extend the analysis of Example 1 in Section 10.5 to the following two exercises:

1. The disk is bounded by a grounded cylindrical vessel of radius b such that $a/b \ll 1$. The disk and the cylinder have the common axis.

2. The disk is placed symmetrically between two grounded parallel plates $z = \pm b$ such that $a/b \ll 1$.

3. Instead of the whole disk, consider the case of an annular disk and extend the analysis of Example 1 of Section 10.5 accordingly. Do the same to the problems in Exercises 1 and 2.

Solve the problems of acoustic diffraction of an axially symmetric plane wave for the configurations in Exercises 4–7.

4. A perfectly rigid circular disk.

5. Perfectly soft and perfectly rigid annular disks.

6. Perfectly soft and rigid spherical caps.

7. Perfectly soft and rigid annular caps.

8. Solve Exercises 1 and 2 when the solid is a spherical cap instead of the circular disk.

For details on mixed boundary value problems, the reader is referred elsewhere [8].

INTEGRAL EQUATION
PERTURBATION METHODS

11.1. BASIC PROCEDURE

In the previous chapter, we solved the Fredholm integral equations of the first kind by converting them to Volterra integral equations and to Fredholm integral equations of the second kind. One of the reasons for the simplicity of that formulation was that we had only one variable of integration. We need to develop methods which solve the integral equations relating to boundaries such as a cylinder or a sphere. In this chapter, we shall deal with three-dimensional problems and present approximate techniques for solving the Fredholm integral equations of the first kind

$$f(P) = \int_S K(P, Q)\, g(Q)\, dS, \qquad P \in S, \tag{1}$$

with $P = \mathbf{x}$ and $Q = \xi$. The analysis for the corresponding plane problems is simpler once the method is grasped.

In the previous chapter, we noticed that certain perturbation parameters naturally arise in physical problems. Let ε be such a parameter occurring in the integral equation (1). Then, we expand all three functions K, f, and g as power series in ε:

$$K = K_0 + \varepsilon K_1 + \varepsilon^2 K_2 + \cdots , \tag{2}$$

$$f = f_0 + \varepsilon f_1 + \varepsilon^2 f_2 + \cdots , \tag{3}$$

$$g = g_0 + \varepsilon g_1 + \varepsilon^2 g_2 + \cdots . \tag{4}$$

Inserting these values in (1) and equating equal powers of ε, we end up in solving the integral equations

$$\int_S K_0 g_0 \, dS = f_0 , \tag{5}$$

$$\int_S K_0 g_1 \, dS = f_1 - \int_S K_1 g_0 \, dS , \tag{6}$$

$$\int_S K_0 g_2 \, dS = f_2 - \int_S K_1 g_1 \, dS - \int_S K_2 g_0 \, dS , \tag{7}$$

and so on.

For the above technique to be useful, the following conditions must be satisfied: (i) $K_0(P, Q)$ is the dominant part of $K(P, Q)$ as in the previous chapter; (ii) the integral equation (5) can be solved; and (iii) the functions $g_0, g_1, \ldots$ are such that the integrals occurring on the right side of equations (6), (7), etc. are easily evaluated.

Fortunately, in this method, it is only the integral equation (5) that needs to be solved because the other integral equations in the sequence have the same kernel.

In certain cases, an approximation to order (ε) can be obtained rather easily. Suppose that the function K_1 occurring in the expansion (2) is a constant A (say), then equation (1) can be written as

$$f(P) + \varepsilon f' = \int_S K_0(P, Q) g(Q) \, dS + O(\varepsilon^2) , \tag{8}$$

where

$$f' = -A \int_S g(Q) \, dS ,$$

is a constant, although as yet unknown. Then, to order ε, equation (8) is similar to the integral equation (5), whose solution is assumed known, and therefore can be solved. The quantity f' can then be evaluated from different considerations. The occurrence of a constant A can be demonstrated by the kernel

$$K(P, Q) = \frac{\exp i\varepsilon |x-\xi|}{|x-\xi|} = \frac{\exp i\varepsilon r}{r} = \frac{1}{r} + i\varepsilon + O(\varepsilon^2) .$$

Here, $K_0(P, Q) = 1/r$ and $A = i$.

In the special case when the kernel K is only of the form $K_0 + A$, the analysis is further simplified. Indeed, suppose that the solution $G(P)$ of the integral equation

$$\int_S K_0(P, Q) G(Q) \, dS = 1 , \qquad P \in S , \tag{9}$$

is known and we are required to solve the equation

$$\int_S [A + K_0(P, Q)] g(Q) \, dS = 1 . \tag{10}$$

We can write (10) in the form

$$\int_S K_0(P, Q) g(Q) \, dS = 1 - A \int_S g(Q) \, dS . \tag{11}$$

Although the value of the integral on the right side of (11) is so far unknown, it is nevertheless a constant, as pointed out earlier. We can therefore divide both sides of equation (11) by the constant factor on the right side and obtain

$$\int_S K_0(P, Q) \left\{ g(Q) / \left[1 - A \int_S g(Q) \, dS \right] \right\} dS = 1 . \tag{12}$$

Comparing (12) with (9), where the value of $G(Q)$ is known, we have

$$g(Q) = \left[1 - A \int_S g(Q) \, dS \right] G(Q) . \tag{13}$$

Integration of this expression over the surface S and a slight rearrangement yields

$$\int_S g(Q) \, dS = \left[\int_S G(Q) \, dS \right] / \left[1 + A \int_S G(Q) \, dS \right] . \tag{14}$$

Subsequently, the substitution of (14) in (13) yields the solution $g(P)$ of the integral equation (10):

$$g(P) = G(P) / \left[1 + A \int_S G(Q) \, dS \right] . \tag{15}$$

This result can be extended to the case when, instead of the constant A in equation (10), we have a separable kernel with finite terms. Suppose again that we know the solution of the integral equations

$$\int_S K_0(P, Q) G_0(Q) \, dS = f(P), \qquad P \in S, \tag{16}$$

and

$$\int_S K_0(P, Q) G_i(Q) \, dS = \psi_i(P); \qquad P \in S, \quad i = 1, \ldots, n; \tag{17}$$

then we can solve the integral equation

$$\int_S [K_0(P, Q) + \sum_{i=1}^n \phi_i(Q) \psi_i(P)] g(Q) \, dS = f(P), \tag{18}$$

where the $\phi_i(Q)$ are known. To accomplish this, we write (18) as

$$\int_S K_0(P, Q) g(Q) \, dS = f(P) - \sum_{i=1}^n C_i \psi_i(P), \qquad P \in S, \tag{19}$$

where

$$C_i = \int_S \phi_i(Q) g(Q) \, dS$$

are constants, as yet unknown. The rest of the steps are a simple repetition of the ones used to solve the integral equation (10).

We now apply these ideas to various disciplines of mathematical physics and engineering. In the sequel, we shall let $G(x; \xi)$ be the generic notation for the Green's function as in Chapters 5 and 6. In view of the expansion (2) for the kernel K, we shall write $G_0(x; \xi)$ for $E(x; \xi)$.

11.2. APPLICATIONS TO ELECTROSTATICS

As a first illustration, we turn to boundary value problems in electrostatics. Let there be two conductors with surfaces S_1 and S_2; S_1 is completely contained in S_2 and is kept at a unit potential, while the potential on S_2 is zero. If a denotes a characteristic length of S_1 and b denotes the minimum distance between a point of S_1 and a point of

S_2, then we have the perturbation parameter $\varepsilon = a/b$, which we assume to be much smaller than unity.

In Section 6.4, we presented an integral representation formula for the electrostatic potential in the region D between S_2 and S_1:

$$f(P) = \int_{S_1} G(P, Q)\sigma(Q)\, dS , \qquad (1)$$

in terms of the Green's function $G(P, Q)$ and the charge density σ. Applying the boundary condition on S_1, equation (1) becomes

$$1 = \int_{S_1} G(P, Q)\sigma(Q)\, dS , \qquad P \in S_1 . \qquad (2)$$

Following the method outlined in the previous section, we write $G(P, Q)$ as the sum of the free-space Green's function $G_0(P, Q)$ and the perturbation term $G_1(P, Q)$ in (2) and get

$$1 = \int_{S_1} G_0(P, Q)\sigma(Q)\, dS + \int_{S_1} G_1(P, Q)\sigma(Q)\, dS , \qquad P \in S_1 . \qquad (3)$$

If the conductor S_2 were absent, we would have only the first integral on the right side of equation (3). Thus, the second integral represents the effect of the conductor S_2 on the potential of S_1. According to the hypothesis of the previous section, we assume that we can solve the integral equation (3) when the second integral is not present.

Note that we can always introduce a constant A and write $G_1 = A + G_2$, where $G_2 = O(A\varepsilon)$. For instance, one possible value of A is the value of $G_1(P, Q)$ for an arbitrary pair of points P and Q on S_1. Hence, the relation (3) can be written as

$$1 = \int_{S_1} G_0(P, Q)\sigma(Q)\, dS + A \int_{S_1} \sigma(Q)\, dS + \int_{S_1} G_2(P, Q)\sigma(Q)\, dS . \qquad (4)$$

Now define a new charge density σ'

$$\sigma'(P) = \sigma(P) \Big/ \Big[1 - A \int_{S_1} \sigma(Q)\, dS\Big] , \qquad (5)$$

from which it follows that

$$\int_{S_1} \sigma\, dS = \Big[\int_{S_1} \sigma'\, dS\Big] \Big/ \Big[[1 + A \int_{S_1} \sigma'\, dS\Big] . \qquad (6)$$

Also, equation (4) can be written in terms of the density σ' as

$$1 = \int_{S_1} G_0(P, Q)\sigma'(Q)\, dS + \int_{S_1} G_2(P, Q)\sigma'(Q)\, dS\,. \tag{7}$$

From the above arguments, we conclude that the second integral on the right side of equation (7) is $O(\varepsilon^2)$ times the first one. If we neglect the terms of this order, then σ' is the electrostatic charge density on S_1 when it is raised to a unit potential in free space. Equation (6) therefore gives the capacity C of the condenser formed by S_1 and S_2 in terms of the free-space capacity C_0 of S_1, that is,

$$C/C_0 = (1 + AC_0)^{-1} + O(\varepsilon^2)\,, \tag{8}$$

or

$$C/C_0 = 1 - AC_0 + O(\varepsilon^2)\,. \tag{9}$$

If A is interpreted as the value of $G_1(P, Q)$ for any pair of points P, Q on S_1, then the result (9) is precisely the capacity that would have been obtained had we used the perturbation procedure (2)–(4). The advantage of equation (8) for determining the electrostatic capacity lies in the fact that in many situations it is possible to show that, by a suitable choice of A, the formula (8) is valid for much higher order in ε.

Example.   We shall elucidate this discussion by the example of a sphere of radius a placed with its center on the axis of an infinite cylinder of radius b. Recall that we gave an integral-equation formulation of a general axially symmetric problem of this nature in Example 2 of Section 6.5. In terms of cylindrical polar coordinates (ρ, φ, z), we found that

$$G_1(\rho, \varphi, z; \rho_1, \varphi_1, z_1) = \sum_{r=0}^{\infty} (2 - \delta_{0r})\, G_1^{(r)}(\rho, z; \rho_1, z_1) \cos r(\varphi - \varphi_1), \tag{10}$$

where

$$G_1^{(r)} = -\frac{2}{\pi} \int_0^{\infty} \frac{I_r(p\rho)\, I_r(p\rho_1)\, K_r(pb)}{I_r(pb)} \cos p(z - z_1)\, dp\,, \tag{11}$$

while the points P and Q are (ρ, φ, z) and (ρ_1, φ_1, z_1), respectively. In view of the axial symmetry, we need only the term $G^{(0)}$ in the above formula.

The next step is to find the constant A in the relation (4). For this purpose, we first set

$$z = a\cos\theta, \qquad \rho = a\sin\theta, \qquad (12)$$

where θ is the angle between Oz and OP, where O is the origin. Then (11) becomes

$$G_1^{(0)} = -\frac{2}{\pi} \operatorname{Re} \int_0^\infty \frac{I_0(pa\sin\theta) I_0(pa\sin\theta_1) K_0(pb)}{I_0(pb)} e^{ipa(\cos\theta - \cos\theta_1)} \, dp, \quad (13)$$

where the "Re" means that we take the real part of the expression. Secondly, we use the formula

$$I_0(pa\sin\theta) e^{ipa\cos\theta} = \sum_{n=0}^\infty \frac{(ipa\sin\theta)^n}{n!} P_n(\cos\theta), \qquad (14)$$

where P_n are the Legendre polynomials. From (13) and (14), it follows that

$$G_1^{(0)}(P, Q) = -\frac{2}{\pi} \operatorname{Re} \int_0^\infty \sum_{m,n=0}^\infty (-1)^m (i)^{m+n} \frac{(pa\sin\theta)^n}{n!} \frac{(pa\sin\theta_1)^m}{m!}$$

$$\times \frac{K_0(pb)}{I_0(pb)} P_n(\cos\theta) P_m(\cos\theta_1) \, dp \qquad (15)$$

$$= -\frac{2}{\pi b} \operatorname{Re} \int_0^\infty (-1)^m (i)^{m+n} \frac{(\varepsilon\sin\theta)^n}{n!} P_n(\cos\theta) \frac{(\varepsilon\sin\theta_1)^m}{m!}$$

$$\times P_m(\cos\theta_1) u^{m+n} \frac{K_0(u)}{I_0(u)} \, du$$

$$= -\frac{1}{b} \operatorname{Re}\left[\sum_{n=0}^\infty \varepsilon^n A_n P_n(\cos\theta) \right]\left[\sum_{n=0}^\infty \varepsilon^n A_n P_n(\cos\theta_1) \right], \quad (16)$$

where $u = pb$ and $\varepsilon = a/b$ (the dimensionless parameter of the problem). Furthermore, the constants A_n are given by the formula

$$A_m A_n = (2/\pi) \int_0^\infty (-1)^m (i)^{m+n} (u)^{m+n} [K_0(u)/I_0(u)] \, du. \qquad (17)$$

It follows from (17) that, when $(m+n)$ is odd, $A_m A_n$ is an imaginary quantity.

The required constant A is now available from (16) and (17):

$$A = -(1/b) A_0{}^2 = -(2/\pi b) \int_0^\infty [K_0(u)/I_0(u)] \, du \, . \tag{18}$$

Substituting this value of A in (8), we have

$$C/C_0 \simeq [1 - (2/\pi b) C_0 I(0)]^{-1} \, , \tag{19}$$

where

$$I(2m) = (2m+1) \int_0^\infty [u^{2m} K_0(u)/I_0(u)] \, du \, , \tag{20}$$

for which numerical tables are available [5].

By careful examination, it can be shown (see Exercise 1) that the above value of the capacity is correct to order ε^6.

11.3. LOW-REYNOLDS-NUMBER HYDRODYNAMICS

Two kinds of linearized equations govern the flow of an incompressible viscous fluid: Stokes and Oseen equations.

Steady Stokes Flow

We have studied these equations in Example 3 of Section 6.7. Recall that, for a free space, the boundary value problem is

$$\nabla^2 \mathbf{q} = \operatorname{grad} p \, , \qquad \operatorname{div} \mathbf{q} = 0 \, ; \tag{1}$$

$$\mathbf{q} = \mathbf{e}_1 \, , \qquad \text{on} \quad S_1 \, ; \qquad q_1 \to 0 \quad \text{at} \quad \infty \, ; \tag{2}$$

where this system has been made dimensionless with the help of the uniform speed U of the solid and with its characteristic length a; here, $\mathbf{e}_1$ is the unit vector along the x_1 axis. The integral-equation formula for this boundary value problem was found in terms of Green's tensor $\mathbf{T}_1$ and Green's vector $\mathbf{p}_1$ to be

$$\mathbf{e}_1 = -\int_{S_1} \mathbf{f} \cdot \mathbf{T}_1 \, dS \,, \qquad P \in S_1 \,, \tag{3}$$

where

$$\mathbf{f} = (\partial \mathbf{q}/\partial n) - p\mathbf{n} \,, \tag{4}$$

$$\mathbf{T}_1 = (1/8\pi)(\mathbf{I} \nabla^2 |\mathbf{x} - \xi| - \operatorname{grad} \operatorname{grad} |\mathbf{x} - \xi|) \,, \tag{5}$$

and

$$\mathbf{p}_1 = -(1/8\pi) \operatorname{grad} \nabla^2 |\mathbf{x} - \xi| \,. \tag{6}$$

The corresponding formula for the resistance $\mathbf{F}_\infty$ (the subscript signifies that we have an infinite mass of fluid) on the body B is found by observing that the stress tensor has the value $\mathbf{I}p + [\nabla \mathbf{q} + (\nabla \mathbf{q})^t]$, where $(\nabla \mathbf{q})^t$ stands for the transpose of $\nabla \mathbf{q}$. Using $(4)_2$, we have

$$\mathbf{F}_\infty = \int_{S_1} \mathbf{f}_\infty \, dS \,. \tag{7}$$

This force can be related to the so-called resistance tensor Φ_∞, which is defined to be such that the force exerted on a body with uniform velocity $\mathbf{u}$ is $\Phi_\infty \cdot \mathbf{u}$. Thus, $\mathbf{F}_\infty = F_\infty \mathbf{e} = -\Phi_\infty \cdot \mathbf{u}$, where $\mathbf{e}$ is the unit vector in the direction of $\mathbf{u}$.

The solutions for various boundary value problems for steady Stokes flow in an unbounded medium are known. As such, the solution of the integral equation (3) can be found for these problems. Hence, the tensor $\mathbf{T}_1$ corresponds to the kernel K_0 of Section 11.1. Below, we shall show how the correction term may be obtained for more complicated cases by using the ideas of Section 11.1. We begin with the boundary effects when the fluid is bounded by a surface S_2.

Boundary Effects on Stokes Flow

The presence of the boundary S_2 necessitates the introduction of a new tensor $\mathbf{T}$ and a corresponding vector $\mathbf{p}$ (see Exercise 7 of Chapter 6). These quantities satisfy the equations

$$\nabla^2 \mathbf{T} - \operatorname{grad} \mathbf{p} = \mathbf{I} \delta(\mathbf{x} - \xi) \,, \qquad \nabla \cdot \mathbf{T} = 0 \qquad \mathbf{T} = 0 \qquad \text{on} \quad S_2 \,. \tag{8}$$

When S_2 tends to infinity $\mathbf{T}$ and $\mathbf{p}$ reduce to $\mathbf{T}_1$ and $\mathbf{p}_1$ as given above. According to the present scheme we write $\mathbf{T} = \mathbf{T}_1 + \mathbf{T}_2$ and $\mathbf{p} = \mathbf{p}_1 + \mathbf{p}_2$ where $\mathbf{T}_2$ and $\mathbf{p}_2$ satisfy the homogeneous part of the system (8).

The integral equation that is equivalent to the present problem is

$$\mathbf{e}_1 = -\int_{S_1} \mathbf{f} \cdot \mathbf{T} \, dS = -\int_{S_1} \mathbf{f} \cdot (\mathbf{T}_1 + \mathbf{T}_2) \, dS . \tag{9}$$

Along with P and Q let us take the origin also on S_1. Furthermore, let ε again be the parameter that gives the ratio of a, the standard geometric length of the solid B, to the minimum distance between a point of S_1 and a point of S_2. Then, by Taylor's theorem, we get

$$\mathbf{T}_2 = \mathbf{T}_2{}^0 + \mathbf{r} \cdot [\text{grad} \, \mathbf{T}_2]_{x=\xi=0} + \xi \cdot [\text{grad}^0 \, \mathbf{T}_2]_{x=\xi=0} + O(\varepsilon^3) , \tag{10}$$

where $\mathbf{T}_2{}^0 = \mathbf{T}_2(0,0)$ and the superscript zero on the grad implies differentiation with respect to the components of ξ. Taking only the first-order terms of the relation (10) in (9), there results the equation

$$\mathbf{e}_1 + \mathbf{F} \cdot \mathbf{T}_2{}^0 = -\int_{S_1} \mathbf{f} \cdot \mathbf{T}_1 \, dS , \tag{11}$$

where $\mathbf{F}$ is defined by relation (7) without the subscript ∞ in that relation, that is, $\mathbf{F}$ is the resistance experienced by B in the bounded medium. The integral equation (11) has the same kernel as that of (3) and, as such, can be considered to give the velocity field in an unbounded fluid when B is moving with uniform velocity $\mathbf{e}_1 + \mathbf{F} \cdot \mathbf{T}_2{}^0$. If we now utilize the concept of the resistance tensor $\mathbf{\Phi}_\infty$ as defined above, we derive the force formula $\mathbf{F}$:

$$\mathbf{F} = -(\mathbf{e}_1 + \mathbf{F} \cdot \mathbf{T}_2{}^0) \cdot \mathbf{\Phi}_\infty . \tag{12}$$

This equation can be solved to give

$$\mathbf{F} = -\mathbf{e}_1 \cdot [\mathbf{\Phi}_\infty^{-1} + \mathbf{T}_2{}^0]^{-1} . \tag{13}$$

Fortunately, replacing $\mathbf{F}$ by $\mathbf{F}_\infty$ introduces error of order ε^2, and thus, to order ε, the formula (12) becomes

$$\mathbf{F} = -(\mathbf{e}_1 + \mathbf{F}_\infty \cdot \mathbf{T}_2{}^0) \cdot \mathbf{\Phi}_\infty . \tag{14}$$

The principal axes of the resistance of B are defined so that, when B moves parallel to one of them in an infinite mass of fluid, the force is in the direction of motion. They are the unit eigenvectors of the resistance tensor Φ_∞. Let us denote them by i_1, i_2, and i_3 such that

$$\Phi_\infty = \Phi_{\infty 1} i_1 i_1 + \Phi_{\infty 2} i_2 i_2 + \Phi_{\infty 3} i_3 i_3 \ . \tag{15}$$

Let us decompose T_2^0 into components with these eigenvectors as the basis. Furthermore, let us set $e_1 = i_1$. Substituting these expressions in the relation (12), we derive

$$F/F_\infty = 1/(1 - \lambda F_\infty) \ , \tag{16}$$

where λ is independent of the form of S_1.

Longitudinal Oscillations of Solids in Stokes Flow

This analysis can be used to obtain an approximate value of the velocity field generated and the resistance experienced by a solid of an arbitrary shape which is executing slow longitudinal vibrations in an unbounded viscous fluid. Let us assume that the body oscillates about some mean position with velocity $Ue^{i\omega t} e_1$ and q and p have the same time dependence. Then, the dimensionless Stokes equations for the steady-state vibrations are

$$-\nabla p + \nabla^2 q - iM^2 q = 0 \ , \qquad \text{div}\, q = 0 \ , \tag{17}$$

where $M^2 = a^2 \omega/\nu$ is the rotational Reynolds number and ν is the coefficient of kinematic viscosity.

The integral representation formulas are the same as for the steady Stokes flow, while T and p now satisfy the equations

$$-\nabla p + \nabla^2 T - iM^2 T = I\delta(x - \xi) \ , \qquad \text{div}\, T = 0 \ , \tag{18}$$

and $T \to 0$ as $x \to \infty$. These equations are satisfied when T and p are given by the formulas

$$T = I\nabla^2 \phi - \text{grad}\,\text{grad}\,\phi \ , \qquad p = -\text{grad}(\nabla^2 - iM^2)\phi \ , \tag{19}$$

$$(\nabla^2 - iM^2)\nabla^2 \phi = \delta(x - \xi) \ . \tag{20}$$

$$\phi = \frac{1 - \exp\{-[(1+i)M/\sqrt{2}]\,|x - \xi|\}}{4\pi iM^2\,|x - \xi|} \ . \tag{21}$$

Thus,

$$\mathbf{T} = \mathbf{T}_1 - [(1+i)/6\pi\sqrt{2}]\,M\mathbf{I} + O(M^2)\,, \tag{22}$$

where $\mathbf{T}_1$ is given by (5).

The next step is to substitute the boundary value $\mathbf{q} = \mathbf{e}_1$ in the integral representation formula for the system (17) and observe that, in view of equation (18) and Green's theorem, we have

$$\int_{S_1} \left(\frac{\partial \mathbf{T}}{\partial n} - \mathbf{pn} \right) dS = -iM^2 \int_{R_i} \mathbf{T}\, dV\,, \tag{23}$$

where R_i is the interior of S_1. The result is the Fredholm integral equation of the first kind for evaluating $\mathbf{f}$:

$$\mathbf{e}_1 - [(1+i)/6\pi\sqrt{2}]\,M\mathbf{F} = -\int_{S_1} \mathbf{T}_1 \cdot \mathbf{f}\, dS + O(M^2)\,, \qquad P \in S_1\,. \tag{24}$$

Following the previous analysis, we have the formula

$$\mathbf{F} = \boldsymbol{\Phi}_\infty \cdot \{\mathbf{e}_1 - [(1+i)/6\pi\sqrt{2}]\,M(\boldsymbol{\Phi}_\infty \cdot \mathbf{e}_1)\} + O(M^2)\,. \tag{25}$$

For a body moving parallel to one of its axes of resistance (which we can take as the x_1 axis of our coordinate system), equation (25) takes the simple form

$$\mathbf{F} = -F_\infty\{1 + [(1+i)/6\pi\sqrt{2}]\,MF_\infty\}\,\mathbf{e}_1\,. \tag{26}$$

For example, for a sphere, $F_\infty = 6\pi\mu aU$ in physical units, where a is the radius of the sphere and μ is the shear viscosity of the fluid. The formula (26) then gives (in physical units)

$$\mathbf{F} = -6\pi\mu aU[1 + (M/\sqrt{2})(1+i)]\,\mathbf{e}_1 + O(M^2)\,. \tag{27}$$

Steady Rotary Stokes Flow

For the rotation of axially symmetric bodies, the pressure is taken to be constant and the steady Stokes equations become

$$\nabla^2 \mathbf{q} = 0\,, \qquad \operatorname{div}\mathbf{q} = 0\,, \qquad p = \text{const}\,. \tag{28}$$

Let the z axis of cylindrical polar coordinates (ρ, φ, z) be the axis of symmetry of these bodies. Assuming that the streamlines are circles lying in planes perpendicular to Oz, then $\mathbf{q}$ has a nonzero component $v(\rho, z)$ in the φ direction only and is independent of φ. The equation of continuity is thus satisfied automatically, while the equation of motion (28) becomes

$$\frac{\partial^2 v}{\partial \rho^2} + \frac{1}{\rho}\frac{\partial v}{\rho \partial} + \frac{\partial^2 v}{\partial z^2} - \frac{v}{\rho^2} = 0, \tag{29}$$

which has been made dimensionless with Ωa as the typical velocity. Here, Ω is the uniform angular velocity of the body and a is its characteristic length. The boundary conditions are

$$v = \rho \quad \text{on} \quad S_1 ; \qquad v = 0 \quad \text{on} \quad S_2 . \tag{30}$$

From (29), it is easily verified that the function

$$w(\rho, \varphi, z) = v(\rho, z)\cos\varphi \tag{31}$$

is harmonic and can therefore be represented in terms of a source density $\sigma(Q)\cos\varphi_1$ spread over S_1, where $\boldsymbol{\xi} = (\rho_1, \varphi_1, z_1)$ are the coordinates of Q and $\sigma(Q)$ is independent of φ_1. Thus, we can use the integral representation formula (6.4.24) for the harmonic function and get

$$w(\rho, \varphi, z) = \int_{S_1} G(P, Q)\sigma(Q)\cos\varphi_1 \, dS , \tag{32}$$

where P is an arbitrary point in the region between S_1 and S_2. On applying the boundary condition $(30)_1$, we obtain

$$\rho = \int_C \rho_1 G^{(1)}(P, Q)\sigma(Q) \, ds , \qquad P \in S_1 , \tag{33}$$

where $\pi^{-1} G^{(1)}(P, Q)$ is the coefficient of $\cos(\varphi - \varphi_1)$ in the Fourier expansion of $G(P, Q)$ and ds denotes the element of the arc length measured along the curve C which is the bounding curve of S_1 in the meridian plane.

Recall the decomposition

$$G(P, Q) = (1/|\mathbf{x} - \boldsymbol{\xi}|) + G_1(P, Q) ,$$

where $G_1(P, Q)$ is finite in the limit as $Q \to P$. We can, similarly, decompose the Fourier component $G^{(1)}$ into the sum

$$G^{(1)} = G_0^{(1)} + G_1^{(1)} ,$$

where $G_0^{(1)}$ arises from the Fourier expansion of $1/|\mathbf{x}-\boldsymbol{\xi}|$ and $G_1^{(1)}$ arises from the expansion of G_1. Therefore, we can write (33) as

$$\rho = \int_C \rho_1 G_0^{(1)} \sigma \, ds + \int_C \rho_1 G_1^{(1)} \sigma \, ds \, . \tag{34}$$

Again let b represent the minimum distance between a point of S_1 and a point of S_2, and we have the small perturbation parameter $\varepsilon = a/b$. The second integral on the right side of (34) is at least of order ε of the first integral. For geometric configurations for which

$$G_1^{(1)} = \rho\rho_1(A + G_2) \, , \tag{35}$$

where A is a constant and G_2 is of order $A\varepsilon$, equation (34) becomes

$$\rho = \int_C \rho_1 G_0^{(1)} \sigma \, ds + A\rho \int_C \rho_1^{\,2} \sigma \, ds + \rho \int_C \rho_1^{\,2} G_2 \sigma \, ds \, , \tag{36}$$

or

$$\rho = \int_C \rho_1 G_0^{(1)} \sigma' \, ds + \rho \int_C \rho_1^{\,2} G_2 \sigma' \, ds \, , \tag{37}$$

where

$$\sigma' = \sigma/(1 - A \int_C \rho_1^{\,2} \sigma \, ds) \, . \tag{38}$$

Consequently, we have the same situation as in the section on electrostatics, that is, σ' represents, with an error which is at most of order ε^2, an appropriate source density for the body rotating in an infinite mass of fluid.

The tangential stress component τ on the surface S_1 in the direction of φ increasing is

$$\tau = \mu\rho \frac{\partial}{\partial n}\left(\frac{v}{\rho}\right) , \tag{39}$$

where $\partial/\partial n$ denotes differentiation along the normal drawn outward to S_1. Furthermore, we know from the analysis of Chapter 6 that the source density $\sigma(Q)$ on S_1 is related to v by

$$4\pi\sigma(Q) = -\rho \frac{\partial}{\partial n}\left(\frac{v}{\rho}\right) . \tag{40}$$

Thus, $\tau = -4\pi\mu\sigma$. From this value of the stress component, the value of the frictional torque N can now be readily calculated to be

$$N = -8\pi^2 \mu \int_C \rho^2 \sigma \, ds \, . \tag{41}$$

The relation between this torque N and the torque N_∞ in an unbounded fluid may be obtained by integrating both sides of the relation (38) around the meridian section C of the axially symmetric body:

$$N = N_\infty [1 + (A/8\pi^2 \mu\Omega) N_\infty]^{-1} \, , \tag{42}$$

with an error of order ε^2. By a suitable choice of A, the formula (42) can be shown to be valid in many cases to a much higher order in ε. Equation (42) can be illustrated with many interesting configurations. For example, the case of a sphere which is symmetrically placed in an infinite cylindrical shell can be studied as in the analysis of Section 11.2. Formula (42) then gives (see Exercise 5)

$$N/N_\infty = [1 + (N_\infty/8\pi\mu\Omega a^3) H_1] \, , \tag{43}$$

where H_k is given by the integral

$$H_k = \frac{2}{\pi} \frac{(-1)^k}{(2k)!} \int_0^\infty x^{2k} \frac{K_1(x)}{I_1(x)} \, dx \, .$$

Rotary Oscillations in Stokes Flow

The equations governing the steady-state rotary oscillations (with circular frequency ω) of axially symmetric solids in an incompressible viscous fluid are

$$(\nabla^2 - iM^2)\mathbf{q} = 0 \, , \qquad \nabla \cdot \mathbf{q} = 0 \, , \tag{44}$$

which are obtained from (17) by setting $p = $ const. As for the steady rotational case above, the only nonzero component of $\mathbf{q}$ is the φ component v, and the differential equations (44) reduce to solving the equation

$$\frac{\partial^2 v}{\partial\rho^2} + \frac{1}{\rho} \frac{\partial v}{\partial\rho} - \frac{v}{\rho^2} + \frac{\partial^2 v}{\partial z^2} - \beta^2 v = 0 \, , \tag{45}$$

where $\beta^2 = iM^2$. We present the analysis for $\beta \ll 1$. The boundary values are

$$v = \rho \quad \text{on} \quad S_1 ; \qquad v = 0 \quad \text{on} \quad S_2 ; \qquad (46)$$

where, as before, S_1 is the surface of the oscillating body and S_2 is the bounding surface.

By writing $w = v \cos \varphi$, equations (45) and (46) reduce to the following boundary value problem:

$$(\nabla^2 - \beta^2) w = 0 , \qquad (47)$$

$$w = \rho \cos \varphi \quad \text{on} \quad S_1 ; \qquad w = 0 \quad \text{on} \quad S_2 . \qquad (48)$$

The Green's function $G(\mathbf{x}; \xi)$ appropriate to this boundary value problem is

$$(\nabla^2 - \beta^2) G(\mathbf{x}; \xi) = -4\pi \delta(\mathbf{x} - \xi) , \qquad G|_{S_2} = 0 . \qquad (49)$$

Thus,

$$G(\mathbf{x}; \xi) = \frac{\exp -\beta |\mathbf{x} - \xi|}{|\mathbf{x} - \xi|} + G_1(\mathbf{x}; \xi) , \qquad (50)$$

where $G_1(\mathbf{x}; \xi)$ is finite in the limit as $\xi \to \mathbf{x}$. The integral representation formula for $w(\mathbf{x})$ follows from Section 6.6:

$$w(\mathbf{x}) = \int_{S_1} \sigma(\rho_1, z_1)(\cos \varphi_1) G(\mathbf{x}; \xi) \, dS , \qquad \xi \in S_1 , \qquad \mathbf{x} \in R , \quad (51)$$

where R is the region between S_1 and S_2, and $\sigma(\rho_1, z_1)$ is given by the formula (40). When we apply the boundary condition $(48)_1$, we obtain the required Fredholm integral equation

$$\rho \cos \varphi = \int_{S_1} \sigma(\rho_1, z_1)(\cos \varphi_1) G(\mathbf{x}; \xi) \, dS , \qquad (52)$$

with $\mathbf{x}$ and ξ on S_1. Now, if $G_1^{(1)}(\rho, z; \rho_1, z_1)$ is the coefficient of $\cos(\varphi - \varphi_1)$ in the Fourier expansion of $G_1(\mathbf{x}; \xi)$, then the integration over φ_1 reduces the above integral to

$$\rho \cos \varphi = \int_{S_1} \sigma(\rho_1, z_1)(\cos \varphi_1) \frac{\exp -\beta |\mathbf{x} - \xi|}{|\mathbf{x} - \xi|} \, dS$$

$$+ \pi(\cos \varphi) \int_C \sigma(\rho_1, z_1) G_1^{(1)}(\rho, z; \rho_1, z_1) \rho_1 \, ds , \quad (53)$$

in the notation of equation (33).

The next step is to expand σ as the perturbation series

$$\sigma = \sum_n \beta^n \sigma_n \tag{54}$$

in equation (53). Moreover, by direct expansion of the Green's function, it can be shown that $G_1^{(1)} = O(\varepsilon^3)$, where ε is the ratio of the characteristic length of the vibrating body to the distance of its center from the nearest point of S_2. It is assumed that $q = \beta/\varepsilon = O(1)$.

Now, equate equal powers of β on both sides of equation (53) and get (after omitting terms that trivially vanish)

$$\rho \cos \varphi = \int_{S_1} \sigma_0(\rho_1, z_1) |\mathbf{x} - \boldsymbol{\xi}|^{-1} \cos \varphi_1 \, dS , \tag{55}$$

$$0 = \int_{S_1} \sigma_1(\rho_1, z_1) |\mathbf{x} - \boldsymbol{\xi}|^{-1} \cos \varphi_1 \, dS , \tag{56}$$

$$0 = \int_{S_1} \sigma_2(\rho_1, z_1) |\mathbf{x} - \boldsymbol{\xi}|^{-1} \cos \varphi_1 \, dS$$
$$+ \tfrac{1}{2} \int_{S_1} \sigma_0(\rho_1, z_1) |\mathbf{x} - \boldsymbol{\xi}| \cos \varphi_1 \, dS , \tag{57}$$

$$0 = \int_{S_1} \sigma_3(\rho_1, z_1) |\mathbf{x} - \boldsymbol{\xi}|^{-1} \cos \varphi \, dS$$
$$+ \tfrac{1}{2} \int \sigma_1(\rho_1, z_1) |\mathbf{x} - \boldsymbol{\xi}| \cos \varphi \, dS$$
$$- \tfrac{1}{6} \int_{S_1} \sigma_0(\rho_1, z_1) |\mathbf{x} - \boldsymbol{\xi}|^2 \cos \varphi \, dS$$
$$+ \pi (\cos \varphi) \int_C \sigma_0(\rho_1, z_1) H(\rho, z; \rho_1, z_1) \rho_1 \, ds , \tag{58}$$

and so on, where

$$G_1^{(1)}(\rho, z; \rho_1, z_1) = \beta^2 H(\rho, z; \rho_1, z_1) + O(\beta^4) . \tag{59}$$

It follows from (55)–(58) that the source densities $\sigma_0, \sigma_1, \sigma_2, \sigma_3$, etc. are determined by solving potential problems in free space of the form encountered in Chapter 6.

The velocity field and the frictional torque can be readily calculated. Indeed, for the evaluation of the torque N, we use the formula (39) and obtain

$$N = \mu \int_S \rho^2 \frac{\partial}{\partial n}\left(\frac{v}{\rho}\right) dS = 2\pi\mu \int_C \rho^3 \frac{\partial}{\partial n}\left(\frac{v}{\rho}\right) ds . \tag{60}$$

From the relations (40), (54), and (60), it follows that

$$N = -8\pi^2 \mu \int_C \rho^2 (\sigma_0 + \beta\sigma_1 + \beta^2\sigma_2 + \beta^3\sigma_3) \, ds + O(\beta^4) . \tag{61}$$

Since potential problems of the type given in equations (55)–(58) can be solved for various configurations such as a sphere, a spheroid, a lens, and a thin circular disk, we can solve our problem for all these geometric shapes. As an example, we consider the case of a thin circular disk vibrating about its axis in a viscous fluid which is contained in an infinite circular cylinder. The axes of the disk and the cylinder coincide. The Green's function for an infinite cylinder $-\infty < z < \infty$, $0 \leqslant \rho \leqslant b$ can be found by following the steps of Example 2, Section 6.5. The result is

$$G(\mathbf{x};\boldsymbol{\xi}) = \frac{\exp -\beta |\mathbf{x}-\boldsymbol{\xi}|}{|\mathbf{x}-\boldsymbol{\xi}|} - \frac{2}{\pi} \sum_{n=0}^{\infty} (2-\delta_{0n}) [\cos n(\varphi - \varphi_1)]$$

$$\times \int_{\beta}^{\infty} \frac{K_n(pb)}{I_n(pb)} I_n(p\rho) I_n(p\rho_1) \{\cos[(p^2 - \beta^2)(z-z_1)]\}$$

$$\times \frac{p \, dp}{(p^2 - \beta^2)^{1/2}} , \tag{62}$$

from which $G_1^{(1)}$ for the disk $\rho \leqslant 1$, $0 \leqslant \varphi \leqslant 2\pi$, $z = 0$ may be readily obtained:

$$G_1^{(1)}(\rho, \rho_1) = -\frac{1}{2\pi} \varepsilon^3 \rho\rho_1 \int_q^{\infty} \frac{K_1(y)}{I_1(y)} \frac{y^3 \, dy}{(y^2 - q^2)^{1/2}} + O(\varepsilon^5) , \tag{63}$$

or

$$H(\rho, \rho_1) = -(1/2\pi q^3)\rho\rho_1 A(q) . \tag{64}$$

Here, $A(q)$ stands for the infinite integral in equation (63).

The integral equations (55)–(58) can be solved by the method of Sections 6.4 and 6.5 (see also Exercise 2 of Chapter 6). The solutions are

$$\sigma_0 = \frac{2\rho}{\pi^2(1-\rho^2)^{1/2}}, \qquad \sigma_1 = 0, \qquad \sigma_2 = \frac{\rho(2-\rho^2)}{3\pi^2(1-\rho^2)^{1/2}},$$

$$\sigma_3 = \frac{4}{3\pi^2}\left[-\frac{2}{3} + \frac{1}{\pi q^3} A(q)\right]\frac{\rho}{(1-\rho^2)^{1/2}}, \qquad 0 \leqslant \rho < 1. \tag{65}$$

Substituting these values in equation (61), we obtain the value of the torque, which in physical units is

$$N = -\frac{32}{3}\,\mu\Omega a^3\left[1 + \frac{1}{5}\beta^2 - \frac{4}{9\pi}\beta^3 + \frac{4}{3\pi^2}\,\varepsilon^3\,A(q)\right]e^{i\omega t} + O(\beta^4, \varepsilon^5). \tag{66}$$

Oseen Flow—Translational Motion

The slow motion past a solid as studied by Oseen is governed by the dimensionless equations (see Example 4, Section 6.7)

$$\mathscr{R}\,\partial\mathbf{q}/\partial x = -\operatorname{grad} p + \nabla^2\mathbf{q}, \qquad \operatorname{div}\mathbf{q} = 0, \tag{67}$$

$$\mathbf{q} = \mathbf{e}_1 \quad \text{on } S_1; \qquad \mathbf{q} = 0 \quad \text{on } S_2. \tag{68}$$

The Fredholm integral equation of the first kind that is equivalent to the boundary value problem (67)–(68) is given by (6.7.52):

$$\mathbf{e}_1 = -\int_S \mathbf{T}\cdot\mathbf{f}\,dS, \tag{69}$$

where the Green's tensor $\mathbf{T}$ and the Green's vector $\mathbf{p}$ are now defined as

$$\left.\begin{array}{l} \mathbf{T} = (1/8\pi)[\mathbf{I}\nabla^2\phi - \operatorname{grad}\operatorname{grad}\phi], \\[2mm] \mathbf{p} = -(1/8\pi)\operatorname{grad}(\nabla^2\phi - \mathscr{R}\,\partial\phi/\partial x_1), \end{array}\right\} \tag{70}$$

$$\phi = (1/|\sigma|)\int_0^{|\sigma|s}[(1-e^{-t})/t]\,dt, \tag{71}$$

and

$$s = |\mathbf{x}-\boldsymbol{\xi}| + (\mathscr{R}/|\mathscr{R}|)(x_1-\xi_1).$$

By using the series

$$\frac{1-e^{-t}}{t} = 1 - \frac{t}{2!} + \frac{t^2}{3!} + \cdots ,$$

we expand ϕ in (71) in terms of the Reynolds number. The relation $(70)_1$ then becomes

$$\mathbf{T} = \mathbf{T}_1 + O(\mathscr{R}) , \tag{72}$$

where $\mathbf{T}_1$ is given by (5). The rest of the analysis is similar to the one given in the subsection above and is left as an exercise for the reader (see Exercise 9).

Oseen Flow—Rotary Motion

By using the present technique, the solutions of the Oseen equations can be presented also for the steady rotations of axially symmetric solids. As in the corresponding Stokes flow case, we take $p = $ const. Then, the Oseen equations take the simple form

$$\mathscr{R}(\partial \mathbf{q}/\partial x_1) - \nabla^2 \mathbf{q} = 0 , \qquad \text{div } \mathbf{q} = 0 . \tag{73}$$

Again, in view of the symmetry, only the φ component V of $\mathbf{q}$ is non-zero and in cylindrical polar coordinates (with $z = x_1$) the boundary value problem becomes

$$\frac{\partial^2 V}{\partial \rho^2} + \frac{1}{\rho} \frac{\partial V}{\partial \rho} - \frac{V}{\rho^2} + \frac{\partial^2 V}{\partial z^2} - 2c \frac{\partial V}{\partial z} = 0 , \tag{74}$$

where $c = Ua/2v = \mathscr{R}/2$. The boundary conditions on V are

$$V = \rho \quad \text{on} \ \ S_1 ; \qquad V = 0 \quad \text{on} \ \ S_2 . \tag{75}$$

The substitution of $V = e^{cz} v(\rho, z)$ reduces this boundary value problem to the following one:

$$\frac{\partial^2 v}{\partial \rho^2} + \frac{1}{\rho} \frac{\partial v}{\partial \rho} - \frac{v}{\rho^2} + \frac{\partial^2 v}{\partial z^2} - c^2 v = 0 , \tag{76}$$

$$v = \rho e^{cz} \quad \text{on} \ \ S_1 ; \qquad v = 0 \quad \text{on} \ \ S_2 . \tag{77}$$

Equation (76) is the same as (45) with β replaced by c. However, the boundary conditions (77) and (46) are different. By repeating the algebraic steps (47)–(53), we end up with the Fredholm integral equation

$$\rho e^{cz} \cos \varphi = \int\limits_{S_1} \sigma(\rho_1, z_1)(\cos \varphi_1)\left[(\exp - c\,|\mathbf{x} - \boldsymbol{\xi}|)/|\mathbf{x} - \boldsymbol{\xi}|\right] dS$$

$$+ \pi(\cos \varphi) \int\limits_{C} \sigma(\rho_1, z_1) G_1^{(1)}(\rho, z; \rho_1, z_1)\rho_1 \, ds \,, \qquad (78)$$

for the evaluation of $\sigma(\rho, z)$ defined by the relation (40). Although the only difference between the integral equations (53) and (78) is in the expression on their left side, it leads to a much more difficult analysis for the present problem. To solve (78), we again take the expansion $\sigma = \sum_n c^n \sigma_n$ as in (54) and also expand $\rho e^{cz} \cos \varphi$ in power series of c. By comparing the equal powers of c in (78), we obtain the following integral equations of potential theory:

$$\rho \cos \varphi = \int\limits_{S_1} \sigma_0(\rho_1, z_1) |\mathbf{x} - \boldsymbol{\xi}|^{-1} \cos \varphi_1 \, dS \,, \qquad (79)$$

$$\rho z \cos \varphi = \int\limits_{S_1} \sigma_1(\rho_1, z_1) |\mathbf{x} - \boldsymbol{\xi}|^{-1} \cos \varphi_1 \, dS \,, \qquad (80)$$

$$\tfrac{1}{2}\rho z^2 \cos \varphi = \int\limits_{S_1} \sigma_2(\rho_1, z_1) |\mathbf{x} - \boldsymbol{\xi}|^{-1} \cos \varphi_1 \, dS$$

$$+ \tfrac{1}{2} \int\limits_{S_1} \sigma_0(\rho_1, z_1) |\mathbf{x} - \boldsymbol{\xi}| \cos \varphi_1 \, dS \,, \qquad (81)$$

$$\tfrac{1}{6}\rho z^3 \cos \varphi = \int\limits_{S_1} \sigma_3(\rho_1, z_1) |\mathbf{x} - \boldsymbol{\xi}|^{-1} \cos \varphi_1 \, dS$$

$$+ \tfrac{1}{2} \int\limits_{S_1} \sigma_1(\rho_1, z_1) |\mathbf{x} - \boldsymbol{\xi}| \cos \varphi_1 \, dS$$

$$- \tfrac{1}{6} \int\limits_{S_1} \sigma_0(\rho_1, z_1) |\mathbf{x} - \boldsymbol{\xi}|^2 \cos \varphi_1 \, dS$$

$$+ \pi(\cos \varphi) \int\limits_{C} \sigma_0(\rho_1, z_1) H(\rho, z; \rho_1, z_1)\rho_1 \, ds \,, \qquad (82)$$

where $H(\rho, z; \rho_1, z_1)$ is defined by relation (59).

For a thin circular disk $z = 0$, $\rho \leqslant 1$, the system of equations (79)–(82) is the same as the system (55)–(58). Thus, the solution for the steady rotation problem for the disk in Oseen flow is the same as the corresponding solution for the steady-state vibrations in Stokes flow. For example, the value of the torque N in the present case can be deduced from the formula (66):

$$N = -\frac{32}{3}\mu\Omega a^3\left\{1 + \frac{c^2}{5} - \frac{4c^3}{9\pi} + \frac{4}{3\pi^2}\,\varepsilon^3 A(q)\right\} + O(c^4,\varepsilon^5), \quad (83)$$

where Ω is the uniform angular velocity of the solid.

For other configurations, one has to solve the integral equations (79)–(82) with nonzero left side. We illustrate this by considering the rotation of a sphere of radius a. In this case, it is convenient to take spherical polar coordinates (r, θ, φ). The value of the Green's function $G(\mathbf{x};\boldsymbol{\xi})$ is the same as (62) with β replaced by c. The corresponding values of $G_1^{(1)}$ and $H(\theta,\theta_1)$ are

$$G_1^{(1)}(\theta,\theta_1) = -\frac{1}{2\pi}\,\varepsilon^3(\sin\theta\sin\theta_1)\int_q^\infty \frac{K_1(y)}{I_1(y)}\frac{y^3\,dy}{(y^2-q^2)^{\frac{1}{2}}} + O(\varepsilon^5), \quad (84)$$

and

$$H(\theta,\theta_1) = -\frac{1}{2\pi q^3}(\sin\theta\sin\theta_1)A(q). \quad (85)$$

The source densities σ_0, σ_1, σ_2, and σ_3 are determined from (79)–(82) by the method of Chapter 6 (see Example 2, Section 6.3, and Exercise 1 of Chapter 6). The result is

$$\sigma_0 = (3/4\pi)P_1^{\,1}(\cos\theta); \qquad \sigma_1 = (5/12\pi)P_2^{\,1}(\cos\theta);$$

$$\sigma_2 = (1/4\pi)[(3/2)P_1^{\,1}(\cos\theta) + (7/15)P_3^{\,1}(\cos\theta)], \quad (86)$$

$$\sigma_3 = -(3/4\pi)P_1^{\,1}(\cos\theta)[(1/3) + (1/2\pi q^3)A(q)],$$

$0 < \theta < \pi$. Substituting these values in the torque formula (61), we obtain (in physical units)

$$N = -8\pi\mu\Omega a^3\left[1 + \frac{4c^2}{15} - \frac{c^3}{3} + \frac{1}{2\pi}\,\varepsilon^3 A(q)\right] + O(c^4,\varepsilon^5). \quad (87)$$

11.4. ELASTICITY

The Navier–Cauchy equations of elasticity are very similar to the equations of Stokes flow and, as such, can be solved rather effectively by this technique. To demonstrate this we first discuss the displacement field in elastostatics.

Elastostatics

The dimensionless equations of elastostatics are

$$(\lambda + \mu)\,\text{grad}\,\theta + \mu \nabla^2 \mathbf{u} = 0\,, \qquad \theta = \text{div}\,\mathbf{u}\,, \tag{1}$$

where $\mathbf{u}$ is the displacement vector and λ and μ are the Lame constants of the medium. The above equations have been made dimensionless by a suitable characteristic length a inherent in the problem. We want to find the displacement field generated by the light, rigid obstacle B with boundary S_1 which is embedded in an unbounded elastic medium and is given a uniform translation d_0 (d_0/a in dimensionless units). Thus, the boundary conditions are

$$\mathbf{u} = (d_0/a)\,\mathbf{e} \qquad \text{on} \quad S_1\,; \qquad \mathbf{u} \to 0 \qquad \text{at} \quad \infty\,. \tag{2}$$

The Fredholm integral equation that is equivalent to the boundary value problem (1)–(2) is (see Exercise 8, Chapter 6)

$$(d_0/a)\mathbf{e} = -\int_{S_1} \mathbf{f} \cdot \mathbf{T}_1 \, dS\,. \tag{3}$$

The Green's tensor $\mathbf{T}_1$ and the dilation vector $\mathbf{\theta}_1$ are defined as

$$T_{1ij} = \frac{1}{8\pi}\left[\frac{\lambda + 3\mu}{\lambda + 2\mu}\frac{\delta_{ij}}{|\mathbf{x} - \boldsymbol{\xi}|} + \frac{\lambda + \mu}{\lambda + 2\mu}\frac{(x_i - \xi_i)(x_j - \xi_j)}{|\mathbf{x} - \boldsymbol{\xi}|^3}\right], \tag{4}$$

$$\theta_{1i} = -\frac{1}{4\pi}\frac{\mu}{\lambda + 2\mu}\frac{x_i - \xi_i}{|\mathbf{x} - \boldsymbol{\xi}|^3}\,, \tag{5}$$

while $\mathbf{f}$ is

$$\mathbf{f} = \mu(d\mathbf{u}/dn) + (\lambda + \mu)\theta\mathbf{n}\,. \tag{6}$$

Since the traction field **t** is defined as

$$t_i = \lambda\theta n_i + \mu n_j(u_{i,j} - u_{j,i}) \,, \tag{7}$$

where by $u_{i,j}$ we mean $\partial u_i/\partial x_j$, we note that the formula for the force **F** acting on the body B is

$$\mathbf{F} = \int_{S_1} \mathbf{f} \, dS \,. \tag{8}$$

As in the cases discussed earlier, formula (3) is the starting point for obtaining the corrections due to the boundary effects as well as the dynamic effects. We assume that the solution of the integral equation (3) is known.

Boundary Effects

Let the elastic medium be bounded by the surface S_2. Within S_2, we define the fundamental tensor **T** and the dilation vector $\boldsymbol{\theta}$ in the same way as $\mathbf{T}_1$ and $\boldsymbol{\theta}_1$. The integral equation corresponding to (3) is

$$(d_0/a)\,\mathbf{e} = -\int_{S_1} \mathbf{f} \cdot \mathbf{T} \, dS \,. \tag{9}$$

Set $\mathbf{T} = \mathbf{T}_1 + \mathbf{T}_2$, where $\mathbf{T}_1$ is given by (4), and $\mathbf{T}_2$, which gives the boundary effects, is regular in the region under consideration.

We now introduce the concept of the traction tensor, which is analogous to the resistance tensor defined in Section 11.3, and denote it by $\boldsymbol{\Phi}_\infty$ for an infinite elastic medium. It has the property that the total static force $\mathbf{F}_\infty$ exerted on a body that has been given a uniform displacement **v** within an infinite elastic medium has the value $-(d_0/a)\boldsymbol{\Phi}_\infty \cdot \mathbf{v}$. For the case of the bounded medium, the corresponding traction tensor is $\boldsymbol{\Phi}$, while the corresponding static force **F** is equivalent to $-(d_0/a)\boldsymbol{\Phi} \cdot \mathbf{v}$. The rest of the analysis is the same as given for the boundary effects on Stokes flow.

Elastodynamics

Here, we derive the dynamical displacement field in an infinite elastic medium in which is embedded a light, rigid body. This body is depressed by an amount $de = d_0 e^{i\omega t} \mathbf{e}$ by an exciting force of the same frequency. The dimensionless steady-state equations of elastodynamics that govern such a motion are

$$[(\lambda + \mu)/\mu] \operatorname{grad} \theta + \nabla^2 \mathbf{u} + m^2 \mathbf{u} = 0 , \qquad (10)$$

where the number $m^2 = \rho_0 \omega^2 a^2/\mu$, and ρ_0 is the density of the medium. Two other numbers also appear in this analysis. They are

$$M^2 = \rho_0 \omega^2 a^2/(\lambda + 2\mu) , \qquad \tau = M/m , \qquad M = O(m) .$$

The integral representation formula for the equation (10) is easily found to be

$$\mathbf{u}(P) = -\int_{S_1} \{ \mathbf{f} \cdot \mathbf{T} - \mathbf{u} \cdot [\mu(d\mathbf{T}/dn) + (\lambda + \mu) \boldsymbol{\theta} \mathbf{n})] \} \, dS , \qquad (11)$$

where $\mathbf{f}$ is defined by (6). The Green's tensor $\mathbf{T}$ and the dilation vector $\boldsymbol{\theta}$ are given by the formulas

$$\mathbf{T} = \frac{1}{8\pi} \left[\mathbf{I}(\nabla^2 + M^2) \phi - \frac{\lambda + \mu}{\lambda + 2\mu} \operatorname{grad} \operatorname{grad} \phi \right] , \qquad (12)$$

$$\boldsymbol{\theta} = \operatorname{div} \mathbf{T} , \qquad (13)$$

where ϕ satisfies the differential equation

$$(\nabla^2 + m^2)(\nabla^2 + M^2) \phi = 8\pi \delta(\mathbf{x} - \boldsymbol{\xi}) . \qquad (14)$$

An appropriate solution of equation (14) is

$$\phi = \frac{2}{m^2 - M^2} \left[\frac{\exp - iM |\mathbf{x} - \boldsymbol{\xi}|}{|\mathbf{x} - \boldsymbol{\xi}|} - \frac{\exp - im |\mathbf{x} - \boldsymbol{\xi}|}{|\mathbf{x} - \boldsymbol{\xi}|} \right] . \qquad (15)$$

For $M \ll 1$, (13) becomes

$$\mathbf{T} = \mathbf{T}_1 - (i/12\pi)(\tau^3 + 2) m\mathbf{I} + O(m^2) , \qquad (16)$$

where $\mathbf{T}_1$ is given by (4).

Since there are two wave velocities involved in this problem, they should satisfy the radiation condition at infinity, while the boundary condition on S_1 is that $\mathbf{u} = (d_0/a)\mathbf{e}$. When we substitute this value in equation (11) and follow the corresponding analysis for the unsteady Stokes flow in the previous section, we obtain the force formula as

$$\mathbf{F} = \boldsymbol{\Phi}_\infty \cdot \left[\frac{d_0}{a} \mathbf{e} - \frac{i}{12\pi} (\tau^3 + 2) m \left(\frac{d_0}{a} \boldsymbol{\Phi}_\infty \cdot \mathbf{e} \right) \right] + O(m^2) . \qquad (17)$$

For axially symmetric solids, (17) takes the simple form

$$\mathbf{F} = -F_\infty \left[1 + \frac{a}{d_0} \frac{F_\infty i}{12\pi} (\tau^3 + 2) m \right] \mathbf{e} + O(m^2) . \qquad (18)$$

The value of the traction tensor $\boldsymbol{\Phi}_\infty$ can be given for various shapes. For example, for an ellipsoid with semiaxes a_1, a_2, and a_3 in the directions of the unit vectors $\mathbf{e}_1$, $\mathbf{e}_2$, and $\mathbf{e}_3$ respectively, we have (in physical units)

$$\boldsymbol{\Phi}_\infty = 16\pi\mu \sum_{i=1}^{3} \left\{ \frac{\mathbf{e}_i \mathbf{e}_i}{(1 - \tau^2) a_i^2 \alpha_i + (1 + \tau^2) \beta} \right\} , \qquad (19)$$

where

$$\alpha_i = \int_0^\infty \frac{d\lambda}{(a_i^2 + \lambda) \Delta(\lambda)} , \qquad \beta = \int_0^\infty \frac{d\lambda}{\Delta(\lambda)} ,$$

$$\Delta^2(\lambda) = (a_1^2 + \lambda)(a_2^2 + \lambda)(a_3^2 + \lambda) .$$

Rotation, Torsion, and Rotary Oscillation Problems in Elasticity

The rotation of axially symmetric inclusions and cavities in an elastic medium are governed by precisely the same partial differential equations as are the rotation problems of Stokes flow as studied in the previous section. We just have to reinterpret some of the symbols. For example, μ now stands for the modulus of rigidity and Ω is the constant angle of rotation.

The low-frequency torsional oscillations of rigid inclusions in a bounded and isotropic elastic medium can be studied by the analysis of the previous section on rotary oscillations in Stokes flow. Indeed, denote the density of the elastic medium by ρ_0, and interpret β as

$$\beta^2 = -\rho_0 \omega^2 a^2 / \mu . \qquad (20)$$

Crack Problems in Elasticity

Let us now consider the problem of an axially symmetric crack with surface S_1 inside an elastic medium bounded by an infinite circular cylinder of radius b. The axes of symmetry of the crack and the cylinder coincide. It is further assumed that the cylinder is maintained under torsion by a torque applied about the axis of symmetry and that S_1 separates the material and hence there is no stress across S_1. The mathematical formulation of this problem is similar to the rotation problem of the steady Stokes flow. As in the previous section, we take cylindrical polar coordinates (ρ, φ, z) with the z axis coincident with the axis of symmetry of S_1. Then, the displacement field has a nonzero component $v(\rho, z)$ in the φ direction only. Similarly, the nonvanishing components of the stresses are

$$\sigma_{z\varphi} = \mu \frac{\partial v}{\partial z}, \qquad \sigma_{\rho\varphi} = \mu \left(\frac{\partial v}{\partial \rho} - \frac{v}{\rho} \right). \tag{21}$$

The equation of equilibrium is

$$\frac{\partial^2 v}{\partial \rho^2} + \frac{1}{\rho} \frac{\partial v}{\partial \rho} - \frac{v}{\rho^2} + \frac{\partial^2 v}{\partial z^2} = 0 . \tag{22}$$

Setting

$$\rho^3 \frac{\partial}{\partial \rho} \left(\frac{v}{\rho} \right) = -\frac{\partial \chi_1}{\partial z} , \qquad \rho^3 \frac{\partial}{\partial z} \left(\frac{v}{\rho} \right) = \frac{\partial \chi_1}{\partial \rho} ,$$

the relations (21) and (22) become

$$\sigma_{z\varphi} = \frac{\mu}{\rho^2} \frac{\partial \chi_1}{\partial \rho} , \qquad \sigma_{\rho\varphi} = -\frac{\mu}{\rho^2} \frac{\partial \chi_1}{\partial z} , \tag{23}$$

and

$$\frac{\partial^2 \chi_1}{\partial \rho^2} - \frac{3}{\rho} \frac{\partial \chi_1}{\partial \rho} + \frac{\partial^2 \chi_1}{\partial z^2} = 0 . \tag{24}$$

Let τ, a constant, denote the angle of twist per unit length for a cylinder without a crack and let the applied torque be $\frac{1}{2}\pi\mu\tau b^2$. Then, by setting

$$\chi_1 = \chi + \tfrac{1}{2}\tau\rho^4 , \qquad \chi = \rho^2 \psi , \tag{25}$$

equation (24) becomes

$$\frac{\partial^2 \psi}{\partial \rho^2} + \frac{1}{\rho}\frac{\partial \psi}{\partial \rho} + \frac{\partial^2 \psi}{\partial z^2} - \frac{4}{\rho^2}\psi = 0 . \tag{26}$$

Note that χ vanishes on the cylinder and is equal to $-\frac{1}{4}\tau\rho^4$ on S_1. Thus, we have the boundary value problem

$$\nabla^2(\psi\cos 2\varphi) = 0 \qquad \text{in} \quad R , \tag{27}$$

$$\psi\cos 2\varphi = -\tfrac{1}{4}\tau\rho^2\cos 2\varphi \qquad \text{on} \quad S_1 , \tag{28}$$

$$\psi\cos 2\varphi = 0 \qquad \text{on} \quad S_2 , \tag{29}$$

where R is the region between S_1 and S_2, while S_2 stands for the surface of the cylinder.

Following the method explained in Sections 6.4, 6.5, and 11.2, we can readily give an integral-equation formulation to this boundary value problem. The result is

$$-\tfrac{1}{4}\tau\rho^2 = \int_C [G_c^{(2)}(\rho, z; \rho_1, z_1) + 2\pi G_1^{(2)}(\rho, z; \rho_1, z_1)]\,\sigma(\rho_1, z_1)\,\rho_1\,ds, \tag{30}$$

where

$$\sigma(\rho_1, z_1) = \frac{1}{4\pi}\left(\frac{\partial \psi}{\partial n}\bigg|_- - \frac{\partial \psi}{\partial n}\bigg|_+\right),$$

and $\pi^{-1}G_0^{(2)}$ is the coefficient of $\cos 2(\varphi - \varphi_1)$ in the Fourier expansion of the free-space Green's function [cf. (6.4.30)] while $G_1^{(2)}$ is defined by equation (11.2.11) with $r = 2$. The curve C is the bounding curve of S_1 in the meridian plane, ds denotes the element of the arc length measured along C, and both (ρ, z) and (ρ_1, z_1) lie on C.

The next step is to introduce a small parameter ε in the problem and also to find the constant A occurring in the relation (11.1.10). Let a denote the maximum distance between two points of C and let $a \ll b$; then, $\varepsilon = a/b$. The constant A is found by substituting the expansion for the modified Bessel function I_2 in the expression for $G_1^{(2)}$. Then, it follows that

$$2\pi G_1^{(2)} = \rho^2\rho_1{}^2[A + G_3(\rho, z; \rho_1, z_1)] , \tag{31}$$

where

$$A = -\frac{3\pi}{64b^5} \beta_4 , \qquad \beta^{2r} = \frac{2^{r+1}}{\pi(2r)!} \int_0^\infty v^{2r} \frac{K_2(v)}{I_2(v)} dv , \qquad (32)$$

and G_3 is of order $A\varepsilon^2$ (see Exercise 14).

Thereafter, the analysis is similar to the one given in Section 11.2. In fact, from (30), we have

$$-\tfrac{1}{4}\tau\rho^2 = \int_C G_0^{(2)} \sigma' \rho_1 \, ds + \rho^2 \int_C G_3 \sigma' \rho_1{}^3 \, ds , \qquad (33)$$

where

$$\sigma' = \sigma[1 + (4A/\tau)\int_C \rho_1{}^3 \sigma \, ds]^{-1} . \qquad (34)$$

The relation (34), in turn, gives

$$\int_C \rho_1{}^3 \sigma \, ds = \int_C \rho_1{}^3 \sigma'[1 - (4A/\tau)\int_C \rho_1{}^3 \sigma' \, ds]^{-1} \, ds . \qquad (35)$$

As in the case of the electrostatic problem of Section 11.2, $\sigma' \simeq \sigma_0$, where σ_0 arises in the integral equation

$$-\tfrac{1}{4}\tau\rho^2 = \int_C G_0^{(2)} \sigma_0 \rho_1 \, ds \qquad (36)$$

for the same crack problem in an infinite medium. Then, it follows from (35) that

$$\int_C \rho_1{}^3 \sigma \, ds \simeq \int_C \rho_1{}^3 \sigma_0[1 - (4A/\tau)\int_C \rho_1{}^3 \sigma_0 \, ds]^{-1} \, ds , \qquad (37)$$

that is, the solution for a crack in a cylinder may be found approximately from the corresponding solution in an infinite medium.

Finally, we evaluate E, the loss of potential energy due to a crack in an infinite cylinder, in terms of E_0, the loss of the corresponding quantity due to a crack in an infinite medium. The value of E is defined by the formula

$$E = (4\pi/\mu) \int_R \rho(\sigma_{z\varphi}^2 + \sigma_{\rho\varphi}^2) \, dV . \qquad (38)$$

Using the relations (23) and (25), and after a slight manipulation, the relation (38) becomes

$$E = -4\pi^2 \mu\tau \int_C \rho^3 \sigma \, ds$$

$$\simeq -4\pi^2 \mu\tau \int_C \rho^3 \sigma_0 \left[1 - (4A/\tau) \int_C \rho_1{}^3 \sigma_0 \, ds\right]^{-1} ds \,, \tag{39}$$

or

$$E \simeq E_0 \left[1 + (AE_0/\mu\tau^2 \pi^2)\right]^{-1} \,. \tag{40}$$

11.5. THEORY OF DIFFRACTION

Finally, we use the method of this chapter for studying the theory of diffraction. Indeed, the method was first introduced for solving problems of this very theory. When we introduce a bounded obstacle B with surface S in a source-free region of the incident field u_i, then this field is disturbed. The total field is $u = u_i + u_s$, where u_s is the diffracted or scattered field, defined only in the exterior region R_e of S. Following the procedure of Section 6.6 and that of Example 1 of Section 6.7, we find that, for a perfectly soft body, we have the following boundary value problem:

$$\nabla^2 k^2 u_s + k^2 u_s = 0 \,, \qquad \mathbf{x} \in R_e \,, \tag{1}$$

$$u_s = -u_i \qquad \text{on} \quad S \,, \tag{2}$$

and u_s satisfies the Sommerfeld radiation condition. The integral-equation representation formula for this problem, by the method and notation of Chapter 6, is

$$u_s(\mathbf{x}) = -\int_S \sigma(\xi) E(\mathbf{x}; \xi) \, dS \,, \tag{3}$$

where $\sigma(\mathbf{x}) = \partial u(\mathbf{x})/\partial n$ is the single-layer density. When $\mathbf{x}$ approaches a point on S, we have, in the limit,

$$u_i(\mathbf{x}) = \int_S \sigma(\xi) E(\mathbf{x}; \xi) \, dS \,, \qquad \mathbf{x}, \xi \in S \,, \tag{4}$$

which is a Fredholm integral equation of the first kind for evaluating the function $\sigma(\mathbf{x})$.

For a perfectly rigid obstacle, we have the Neumann boundary value problem consisting of the differential equation (1) and the boundary condition

$$\partial u_s/\partial n = -\partial u_i/\partial n \qquad \text{on} \quad S. \tag{5}$$

The integral representation formula that embodies (1) and (5) is

$$u_s(\mathbf{x}) = \int_S u(\xi)\,[\partial E(\mathbf{x};\xi)/\partial n]\,dS, \qquad \mathbf{x} \in R_e. \tag{6}$$

When $\mathbf{x}$ approaches a point on S, we use the relation (6.6.10) and obtain

$$u_s(\mathbf{x}) = \tfrac{1}{2}u(\mathbf{x}) + \int_S u(\xi)\,[\partial E(\mathbf{x};\xi)/\partial n]\,dS, \qquad \mathbf{x}, \xi \in S \tag{7}$$

or

$$2u_i(\mathbf{x}) = \tau(\mathbf{x}) - 2\int_S \tau(\xi)\,[\partial E(\mathbf{x};\xi)/\partial n]\,dS, \qquad \mathbf{x}, \xi \in S, \tag{8}$$

where $\tau(\mathbf{x}) = u(\mathbf{x})$ is the double-layer density.

Example.   Let us solve the problem of the diffraction of a plane wave by a soft sphere. It is convenient to take spherical polar coordinates (r, θ, φ). Thus, $\mathbf{x} = P = (r, \theta, \varphi)$, $\xi = Q = (a, \theta_1, \varphi_1)$, and $|\mathbf{x}-\xi| = R$.

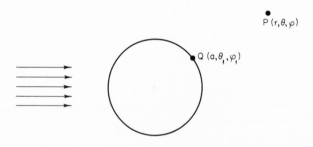

Figure 11.1

In this case (see Figure 11.1)

$$u_i(P) = e^{iak(\cos\theta)} = 1 + iak(\cos\theta) - \frac{a^2 k^2}{2}(\cos^2\theta)$$

$$- i\frac{a^3 k^3 \cos^3\theta}{6} + O(k^4) . \tag{9}$$

$$E(P;Q) = \frac{e^{ikR}}{4\pi R} = \frac{1}{4\pi R} + \frac{ik}{4\pi} - \frac{k^2 R}{8\pi} - \frac{ik^3 R^2}{24\pi} + O(k^4). \tag{10}$$

$$\sigma(Q) = \sigma_0(Q) + k\sigma_1(Q) + k^2\sigma_2(Q) + k^3\sigma_3(Q) + O(k^4). \tag{11}$$

Substituting (9)–(11) in (4) and equating equal powers of k, we have

$$1 = \int_S \frac{\sigma_0(Q)}{4\pi R} \, dS , \tag{12}$$

$$ia\cos\theta = \int_S \frac{\sigma_1(Q)}{4\pi R} \, dS + \int_S \frac{i\sigma_0(Q)}{4\pi} \, dS , \tag{13}$$

$$-\frac{a^2}{2}\cos^2\theta = \int_S \frac{\sigma_2(Q)}{4\pi R} \, dS + \int_S \frac{i\sigma_1(Q)}{4\pi} \, dS - \int_S \frac{R\sigma_0(Q)}{8\pi} \, dS , \tag{14}$$

and so on. The integral equations (12)–(14) are of the same form as (11.3.79)–(11.3.82) and are easily solved by introducing Legendre polynomials. The solution is

$$\sigma_0 = 1/a , \qquad \sigma_1 = -i + 3iP_1(\cos\theta_1) = -i + 3i\cos\theta_1 ,$$
$$\sigma_2 = -\tfrac{1}{2}aP_0(\cos\theta_1) - \tfrac{5}{3}aP_2(\cos\theta_1) = \tfrac{1}{3}a - \tfrac{5}{2}a\cos^2\theta_1 . \tag{15}$$

Consequently,

$$\sigma(Q) = \sigma(\theta_1)$$

$$= (1/a)\left[1 + i\varepsilon(3\cos\theta_1 - 1) + \varepsilon^2(\tfrac{1}{3} - \tfrac{5}{2}\cos^2\theta_1) + O(\varepsilon^3)\right] , \tag{16}$$

where $\varepsilon = ak$ and we assume that $\varepsilon \ll 1$.

The quantities of physical interest in the theory of scattering are the far-field amplitude and the scattering cross section, which we now evaluate for the present example. Since the scattered wave is an outgoing wave and satisfies the radiation condition, we expect its far-field behavior to be given by

$$u_s(r, \theta, \varphi) \simeq A(\theta, \varphi) e^{ikr}/r . \tag{17}$$

Now, for large r,

$$R = |\mathbf{x} - \boldsymbol{\xi}| = (r^2 + a^2 - 2ar \cos \gamma)^{1/2}$$

$$= r[1 + (a^2/r^2) - (2a/r) \cos \gamma]^{1/2} \simeq (r - a \cos \gamma) , \tag{18}$$

where

$$\cos \gamma = \cos \theta \cos \theta_1 + \sin \theta \sin \theta_1 \cos(\varphi - \varphi_1) .$$

Hence, from (3), we have

$$u_s(r, \theta, \varphi) \simeq A(\theta) e^{ikr}/r , \tag{19}$$

where

$$A(\theta) = - (a^2/4\pi) \int_0^\pi \int_0^{2\pi} e^{-i\varepsilon \cos \gamma} \sigma(\theta_1) \sin \theta_1 \, d\theta_1 \, d\varphi_1 , \tag{20}$$

is the required far-field amplitude. The next step is to use the expansions of the terms inside the integral sign of (20), integrate, and obtain

$$A(\theta) = - a[1 - i\varepsilon + \varepsilon^2(\cos \theta - \tfrac{2}{3}) + O(\varepsilon^3)] . \tag{21}$$

The value of the scattering cross section is given by the formula

$$\text{S.C.S.} = 2\pi \int_0^\pi |A(\theta)|^2 \sin \theta \, d\theta = 4\pi a^2 [1 - (\varepsilon^2/3) + O(\varepsilon^4)] . \tag{22}$$

EXERCISES

1. Show that formula (11.2.19) for the capacity is correct to order ε^6.

2. Instead of spherical coordinates (11.2.12), take oblate-spheroidal coordinates

$$z = ae\xi , \qquad \rho = ae[(1 - \xi^2)(1 + \zeta^2)]^{1/2} ,$$

and solve the electrostatic potential problem for the case of an oblate spheroid placed symmetrically inside a grounded infinite cylinder of radius b. Show that the capacity C of this condenser is

$$C = \frac{ae}{\sin^{-1}e}\left\{1 - \frac{2\varepsilon e}{\pi \sin^{-1}e}\right.$$

$$\times \left[I(0) + \frac{1}{9}(e\varepsilon)^2\frac{I(2)}{9} + \frac{2}{225}(e\varepsilon)^4 I(4)\right]\right\}^{-1} + O(\varepsilon^6),$$

where the quantities $I(2m)$ are defined by (11.2.20).

Hint: Use the expansion

$$I_0(p\rho)e^{ipz} = \sum B_n(p)P_n(\xi)P_n(i\zeta),$$

where the P_n are Legendre polynomials.

3. Following a procedure similar to that in Exercise 2, solve the electrostatic potential problem for a prolate spheroid.

4. Solve Exercises 2 and 3 when the spheroids are placed between two grounded parallel plates.

5. Derive formula (11.3.43).

6. Substitute the values of the densities σ_0, σ_1, σ_2, and σ_3 as given by (11.3.65) in (11.3.51) and obtain the velocity field.

7. By using formula (11.3.61), obtain the frictional torque experienced by a sphere of radius a which is oscillating in a viscous fluid.

8. Do the same as in Exercise 7 for a spheroid and a disk.

9. Calculate the $O(\mathscr{R})$ term in formula (11.3.72) explicitly and use this expression to deduce an approximate formula for the drag experienced by a solid in Oseen flow. Illustrate this formula for the case of a sphere.

10. Use the values of the charge densities as given by the formulas (11.3.86) and evaluate the velocity field set up in Oseen flow when a sphere is rotating uniformly and is placed symmetrically inside a circular cylinder.

11. Find the torque experienced by a sphere which is rotating uniformly in Oseen flow and is bounded by a pair of parallel walls $z = \pm c$. Evaluate also the velocity field.

12. Extend the analysis of the steady Oseen flow in the text to the case of the steady-state vibrations of axially symmetric solids in Oseen flow.

13. Use the analysis of Section 11.4 and discuss the problem of a spherical inclusion in an elastic half-space.

14. Derive the integral equation (11.4.30). Prove that the quantity G_3 occurring in the relation (11.4.31) is of order $A\varepsilon^2$. Also prove that σ' occurring in the relation (11.4.33) is such that $\sigma' = \sigma_0[1 + O(\varepsilon^7)]$.

15. Start with the integral equation (11.5.8) and solve the problem of diffraction of a plane wave by a perfectly rigid sphere.

16. Solve the problem of diffraction of a plane wave by a perfectly soft and by a rigid circular disk. Also solve the dual problem of diffraction by an aperture.

17. Use the approximation

$$\frac{i}{4} H_0^{(1)}(kR) = -\frac{1}{2\pi}(q + \log R) + \frac{k^2 R^2}{8\pi}(q - 1 + \log R) + \log \tfrac{1}{2}k \{O(k^4)\},$$

where $q = \gamma + \log \tfrac{1}{2}k - \tfrac{1}{2}\pi i$ (γ is Euler's constant), and solve the problem of diffraction of a plane wave by a soft and by a rigid circular cylinder.

APPENDIX

A.1. To prove the identity

$$\int_a^s \int_a^{s_n} \cdots \int_a^{s_3} \int_a^{s_2} F(s_1) \, ds_1 \, ds_2 \cdots ds_{n-1} \, ds_n$$

$$= [1/(n-1)!] \int_a^s (s-t)^{n-1} \; F(t) \, dt \,, \tag{1}$$

we begin with the formula

$$\frac{d}{ds} \int_{A(s)}^{B(s)} f(s,t) \, dt = \int_A^B \frac{\partial f(s,t)}{\partial s} \, dt + f[s, B(s)] \frac{dB}{ds} - f[s, A(s)] \frac{dA}{ds} \,, \tag{2}$$

for differentiation of an integral involving a parameter. Let us apply this formula for the differentiation of the function $I_n(s)$ defined by the relation

$$I_n(s) = \int_a^s (s-t)^{n-1} F(t) \, dt \,, \tag{3}$$

285

where n is a positive integer and a is a constant. For this purpose, set

$$f(s,t) = (s-t)^{n-1} F(t)$$

in (2). The result is

$$dI_n/ds = (n-1) \int_a^s (s-t)^{n-2} F(t)\, dt + [(s-t)^{n-1} F(t)]_{t=s}$$

$$= (n-1) I_{n-1}, \qquad n > 1. \tag{4}$$

For $n = 1$, we have directly from (3)

$$dI_1/ds = F(s). \tag{5}$$

From the recurrence relation (4), we obtain

$$d^k I_n/ds^k = (n-1)(n-2)\cdots(n-k) I_{n-k}, \qquad n > k, \tag{6}$$

which for $k = n-1$ becomes

$$d^{n-1} I_n/ds^{n-1} = (n-1)!\, I_1(s). \tag{7}$$

Differentiating (7) and using (5) results in the equation

$$d^n I_n/ds^n = (n-1)!\, F(s). \tag{8}$$

Furthermore, from the relations (3), (6), and (7) it follows that $I_n(s)$ and its first $n-1$ derivatives all vanish when $s = a$. Hence, equations (5) and (8) yield

$$I_1(s) = \int_a^s F(s_1)\, ds_1,$$

$$I_2(s) = \int_a^s I_1(s_2)\, ds_2 = \int_a^s \int_a^{s_2} F(s_1)\, ds_1\, ds_2, \tag{9}$$

$$I_n(s) = (n-1)! \int_a^s \int_a^{s_n} \cdots \int_a^{s_3} \int_a^{s_2} F(s_1)\, ds_1\, ds_2 \cdots ds_{n-1}\, ds_n. \tag{10}$$

Combining (3) and (10), we have the required identity (1).

A.2. It can be easily proved by the complex integration method that

$$\int_0^\infty p\left(\frac{p}{\gamma}-1\right)J_\mu(pv)J_\mu(p\rho)\,dp = \begin{cases} i\int_0^k \dfrac{p^2}{(k^2-p^2)^{1/2}}\,H_\mu^{(1)}(pv)J_\mu(p\rho)\,dp\,, \\[4pt] \qquad v\geqslant\rho\,, \\[16pt] i\int_0^k \dfrac{p^2}{(k^2-p^2)^{1/2}}\,J_\mu(pv)\,H_\mu^{(1)}(p\rho)\,dp\,, \\[4pt] \qquad \rho\geqslant v\,, \end{cases} \qquad (11)$$

where

$$\gamma = \begin{cases} -i(k^2-p^2)^{1/2}\,, & k\geqslant p\,, \\ (p^2-k^2)^{1/2}\,, & p\geqslant k\,, \end{cases} \qquad (12)$$

and $v,\rho>0$.

Suppose that $v\geqslant\rho$ and let the complex plane be $s=\sigma+i\tau$. Integrating

$$\{[s^2/(s^2-k^2)^{1/2}]-s\}\,H_\mu^{(1)}(sv)J_\mu(s\rho)$$

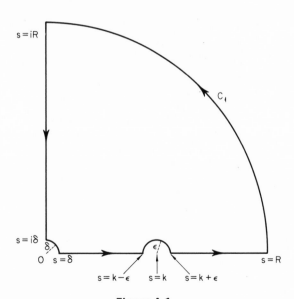

Figure A.1

around a contour C_1 in the upper right-hand quadrant passing over the branch point $s = k$, as shown in Figure A.1, we get

$$\oint_{C_1} \{[s^2/(s^2 - k^2)^{1/2}] - s\} H_\mu^{(1)}(sv) J_\mu(s\rho) \, ds = 0 \,,$$

because there are no singularities within this contour. If we let $\delta, \varepsilon \to 0$ and $R \to \infty$, the contributions from the corresponding arcs tend to zero. Hence,

$$\int_0^k \left(\frac{\sigma^2}{i(k^2 - \sigma^2)^{1/2}} - \sigma \right) H_\mu^{(1)}(\sigma v) J_\mu(\sigma\rho) \, d\sigma + \int_k^\infty \left(\frac{\sigma^2}{(\sigma^2 - k^2)^{1/2}} - \sigma \right)$$

$$\times H_\mu^{(1)}(\sigma v) J_\mu(\sigma\rho) \, d\sigma$$

$$+ i \int_\infty^0 \left(\frac{-\tau^2}{i(\tau^2 + k^2)^{1/2}} - i\tau \right) H_\mu^{(1)}(i\tau v) J_\mu(i\tau\rho) \, d\tau = 0 \,. \tag{13}$$

Similarly, integrating

$$\{[s^2/(s^2 - k^2)^{1/2}] - s\} H_\mu^{(2)}(sv) J_\mu(s\rho)$$

around a contour C_2 in the lower right-hand quadrant and passing under the branch point $s = k$ gives

$$\int_0^k \left(\frac{\sigma^2}{-i(k^2 - \sigma^2)^{1/2}} - \sigma \right) H_\mu^{(2)}(\sigma v) J_\mu(\sigma\rho) \, d\sigma + \int_k^\infty \left(\frac{\sigma^2}{(\sigma^2 - k^2)^{1/2}} - \sigma \right)$$

$$\times H_\mu^{(2)}(\sigma v) J_\mu(\sigma\rho) \, d\sigma$$

$$- i \int_\infty^0 \left(\frac{\tau^2}{i(\tau^2 + \sigma^2)} + i\tau \right) H_\mu^{(2)}(-i\tau v) J_\mu(-i\tau\rho) \, d\tau = 0 \,. \tag{14}$$

Now using the relation $H_\mu^{(1)}(i\tau v) J_\mu(i\tau\rho) = -H_\mu^{(2)}(-i\tau v) J_\mu(-i\tau\rho)$ and adding equations (13) and (14), we obtain, for $v \geqslant \rho$,

$$\int\limits_{k}^{\infty}\left(\frac{\sigma^2}{(\sigma^2-k^2)^{\frac{1}{2}}}-\sigma\right)J_\mu(\sigma v)J_\mu(\sigma\rho)\,d\sigma$$

$$=\int\limits_{0}^{k}\sigma J_\mu(\sigma v)J_\mu(\sigma\rho)\,d\sigma-\int\limits_{0}^{k}\frac{\sigma^2}{(k^2-\sigma^2)^{\frac{1}{2}}}\,Y_\mu(\sigma v)J_\mu(\sigma\rho)\,d\sigma,\quad(15)$$

where $Y\mu$ is the Bessel function of the second kind. It follows from (13) and (15) that the left-hand side of (11) is

$$\int\limits_{0}^{\infty}p\left(\frac{p}{\gamma}-1\right)J_\mu(pv)J_\mu(p\rho)\,dp$$

$$=\int\limits_{0}^{k}\left(\frac{ip^2}{(k^2-p^2)^{\frac{1}{2}}}-p\right)J_\mu(pv)J_\mu(p\rho)\,dp$$

$$+\int\limits_{k}^{\infty}\left(\frac{p^2}{(p^2-k^2)^{\frac{1}{2}}}-p\right)J_\mu(pv)J_\mu(p\rho)\,dp$$

$$=i\int\limits_{0}^{k}\frac{p^2}{(k^2-p^2)^{\frac{1}{2}}}\,[J_\mu(pv)+iY_\mu(pv)]J_\mu(p\rho)\,dp$$

$$=i\int\limits_{0}^{k}\frac{p^2}{(p^2-k^2)^{\frac{1}{2}}}\,H_\mu^{(1)}(pv)J_\mu(p\rho)\,dp,\qquad v\geqslant\rho,$$

which proves the first part of formula (11). The second part follows in a similar fashion.

BIBLIOGRAPHY

1. Bocher, M., "An Introduction to the Study of Integral Equations" (Cambridge Tracts in Mathematics and Mathematical Physics, No. 10). Cambridge Univ. Press, 1909.
2. Bückner, H., "Die practische Behandlung von Integralgleichungen" (Ergebnisse der angewandten Mathematik, No. 1). Springer, New York, 1952.
3. Carrier, G. F., Krook, M., and Pearson, C. F., "Functions of a Complex Variable." McGraw-Hill, New York, 1966.
4. Courant, R., and Hilbert, D., "Methods of Mathematical Physics," Vols. I and II. Wiley, New York, 1953, 1962.
5. Haberman, W. L., and Harley, E. E., Numerical evaluation of integrals containing modified Bessel functions, David Taylor Model Basin Report No. 1580, 1964.
6. Hilderbrand, F. B., "Methods of Applied Mathematics," 2nd ed. Prentice-Hall, Englewood, New Jersey, 1965.
7. Irving, J., and Mullineux, N., "Methods in Physics and Engineering." Academic Press, New York, 1959.
8. Jain, D. L., and Kanwal, R. P., "Mixed Boundary Value Problems of Mathematical Physics." In preparation.
9. Kellog, O. D., "Foundations of Potential Theory." Dover, New York, 1953.
10. Lovitt, W. V., "Linear Integral Equations." Dover, New York, 1950.
11. Mikhlin, S. G., "Integral Equations." Pergamon Press, Oxford, 1957.
12. Morse, P. M., and Feshback, H., "Methods of Theoretical Physics," Vols. I and II. McGraw-Hill, New York, 1953.

290

13. Mushkhelishvili, N. I., "Singular Integral Equations," 2nd ed. P. Noordhoff N.V., Gronignen, Holland, 1946.
14. Peterovskii, I. G., "Integral Equations," Grylock Press, New York, 1957.
15. Pogorzelski, W., "Integral Equations and Their Applications," Vol. 1. Pergamon Press, Oxford, 1966.
16. Smirnov, V. I., "Integral Equations and Partial Differential Equations." Addison-Wesley, Reading, Massachusetts, 1964.
17. Sneddon, I. N., "Fourier Transforms." McGraw-Hill, New York, 1951.
18. Sneddon, I. N., "Mixed Boundary Value Problems in Potential Theory." Wiley, New York, 1966.
19. Stakgold, I., "Boundary Value Problems of Mathematical Physics," Vols. I and II. Macmillan, New York, 1967, 1968.
20. Tricomi, F. G., "Integral Equations." Wiley, New York, 1957.
21. Watson, G. N., "A Treatise on the Theory of Bessel Functions." Cambridge Univ. Press, London and New York, 1962.
22. Weinberger, H. F., "A First Course in Partial Differential Equations." Blaisdell, New York, 1965.
23. Yosida, K., "Lectures on Differential and Integral Equations." Wiley, New York, 1960.

INDEX